MICROWAVE COOKING AND PROCESSING

Engineering Fundamentals for the Food Scientist

MICROWAVE COOKING AND PROCESSING

Engineering Fundamentals for the Food Scientist

Charles R. Buffler, PhD

Kraft General Foods
Technical Center
Glenview, Illinois

An avi Book
Published by Van Nostrand Reinhold
New York

An AVI Book
(AVI is an imprint of Van Nostrand Reinhold)
Copyright © 1993 by Van Nostrand Reinhold
Library of Congress Catalog Card Number 92-19297
ISBN 0-442-00867-8

Printed in the United States of America.

Van Nostrand Reinhold
115 Fifth Avenue
New York, New York 10003

Chapman and Hall
2-6 Boundary Row
London, SE1 8HN, England

Thomas Nelson Australia
102 Dodds Street
South Melbourne 3205
Victoria, Australia

Nelson Canada
1120 Birchmount Road
Scarborough, Ontario MIK 5G4, Canada

16 15 14 13 12 11 10 9 8 7 6 5 4 3 2 1

Library of Congress Cataloging-in-Publication Data
Buffler, Charles R.
    Microwave cooking and processing : engineering fundamentals for
the food scientist / Charles R. Buffler.
        p. cm.
    "An AVI book."
    Includes bibliographical references and index.
    ISBN 0-442-00867-8
    1. Microwave ovens. 2. Microwave cookery. I. Title.
TX657.064B84 1992
641.5'882—dc20   92-19297
                                                         CIP

Contents

Preface

The general public first started taking notice of the microwave oven in 1975 when sales volume, for the first time, exceeded 1,000,000 units per year! In 1985, ovens were owned by over 50% of U.S. households, and food companies were developing microwavable products in earnest. During this period microwave product developers found that very little information was available to assist them with their task. Microwave ovens utilized a technology that was completely different from conventional and well-understood cooking procedures. The interaction of microwaves with foods behaved in a mysterious fashion, and product development was, at best, guesswork and, at least, trial and error. This lack of knowledge of both the microwave oven and microwave-food interaction spurred the development of this text.

Microwave technology was developed during World War II when the best scientists in the United States were cloistered at the radiation laboratory at MIT. In the 1940s virtually all the technology that could impact the microwave oven was developed and well understood. The problem was in the translation of this decades-old knowledge to rules and techniques to be used for food product development. Early authors such as Püschner and Copson wrote texts primarily on applications to industrial processing. This knowledge lay fallow as these texts went out of print. In 1973, Helen Van Zante published her text, *The Microwave Oven,* which for the first time addressed the technical issues of the oven itself as well as the arcane aspects of microwave cooking. Unfortunately this text also went out of print.

As was evident by attendance at many microwave oven technology courses held throughout the world over the past 10 years, there was a need for a codified body of technical knowledge expanding the pioneer work of Van Zante. In 1989, while corresponding with her, I was inspired by her suggestion to collaborate with her on a second edition. Unfortunately we were unable to commence the project because of her death.

Because of the continuing need for the information, and with the help of my sponsoring editor, Dr. Eleanor Riemer, the format for this book was developed. The text is designed as a teaching tool—a mechanism to understand the various ramifications of the microwave oven, microwave interaction with foods and materials, special safety issues, and microwave processing. It is designed for researchers in various fields who need to make microwave technology an adjunct to their specialty, but who may not wish to make microwave technology their main focus. Specifically, the text is designed for food scientists, food lab technologists, chemists, nutrition specialists, packaging engineers, recipe and package direction developers, consumer affairs and public relations personnel (including 800 operators), and plant and safety engineers.

The text is designed to be a translation from electrical engineering to a language that can be understood and used by researchers in other fields. For this reason, concepts may be oversimplified, but are essentially correct. Technical purists may object, but the intent is to provide a feel for what is happening in the oven and to the food during the microwaving process. How many times does a food scientist or other researcher attend a lecture by a technical specialist and come away understanding nothing? While the book is designed with a thrust toward the food scientist and food product developer, virtually every portion is appropriate to any field requiring knowledge of microwave technology. The engineering fundamentals are the same. The same concepts apply to the study of the radiation absorbed by a human being from a television broadcast station as to the cooking of a roast in a microwave oven.

Scientific (SI) units are used extensively. In some instances, units in common use are also included. For example, in Chapter 10, English units are used since little metrification has occurred in the U.S. processing industry. When temperature is discussed, units are given in both Celsius (°C) and Fahrenheit (°F). In formulas when SI units are given, K is used. For daily work °C may be used because the temperatures in question are in terms of changes in temperature not absolute values.

I would like to acknowledge the contribution to my personal body of knowledge that I have gained over the years from my colleagues, Dr. Ronald Lentz, Professor Richard Mudgett, Per Risman, Harry Rubbright, Bob Schiffmann, and Dr. Hans Steyskal. In addition to their contribution, a special debt of thanks is owed to my colleagues who helped me with editing and correcting portions of this manuscript: Dr. Aaron Brody, Dr. Robert Decareau, Joy Daniel, Dr. John Humber, Dr. Claude Lorenson, Dr. John Osepchuk, Dr. Pete Snyder, Dr. Marlene Stanford, and Glen Walters. Also thanks are due to Phil Litsas and Keith Cerosky, whose technical assistance was invaluable in performing the research quoted in numerous of my references. I would also like to acknowledge the excellent work of Bill Morello for preparing all the line drawings. Finally, a great deal of appreciation to my wife Nancy for her continued support and her patience in putting up with the incessant whirring of the word processor for the multitudinous months it took to prepare the manuscript.

Charles R. Buffler
Marlborough, New Hampshire

Symbols and Abbreviations

A	fractional amount of ash in sample
A	cavity dimension in x direction (width)
am	amplitude modulated (modulation)
ASTM	American Society of Testing and Materials
B	cavity dimension in y direction (height)
Btu/lb	British thermal units per pound
Btu/lb°F	British thermal units per pound degree Fahrenheit
Btu/kWh	British thermal units per kilowatt hour
c	velocity (speed) of light and electromagnetic waves
C	cavity dimension in z direction (depth)
C	fractional amount of carbohydrate in sample
C_1, C_2	capacitor designation
C_p	heat capacity
CPET	crystallized polyester
cal/min	calories per minute
cal/g°C	calories per gram degree Celsius
cm	centimeters
cm/s	centimeters per second
d	distance
d_p	power penetration depth (36.8%)
D(E/e)	electric field penetration depth (36.8%)
D(P/e)	power penetration depth (36.8%)
D(P/2)	one half power penetration depth (50%)

DOE	Department of Energy
e	Napierian logarithm base; 2.718
E	electric field
e$_{ff}$	efficiency of microwave system
E$_{ff}$	process efficiency
EVOH	ethylene vinyl alcohol
f	frequency
f	fractional weight lost
f$_c$	critical frequency for relaxation
FCC	Federal Communications Commission
FDA	Food and Drug Administration
FD-TD	finite difference-time domain
FEM	finite element method
g	grams
g/cm^3	grams per cubic centimeter
g/mL	grams per milliliter
gms	grams
F	fractional amount of fat in sample
fm	frequency modulated (modulation)
ft^3	cubic feet
GHz	gigahertz
H	magnetic field
Hz	hertz (cycles per second)
IMPI	International Microwave Power Institute
in	inches
in/ns	inches per nanosecond
IEC	International Electrotechnical Commission
ISM	industrial, scientific and medical
J/kg	joules per kilogram
K	absolute temperature (kelvin)
kcal/kg	kilocalories per kilogram
kg	kilogram
kg/m^3	kilograms per cubic meter
kHz	kilohertz
kJ	kilojouls
kJ/g	kilojouls per gram

kW	kilowatt
kWh	kilowatt hour
l	integer index indicating number of maxima in standing wave in x direction
L	liter
lb	pound
lb/h	pound per hour
m	integer index indicating number of maxima in standing wave in y direction
m	meters
m/ns	meters per nanosecond
m/s	meters per second
m²/s	square meters per second
m³	cubic meters
MHz	megahertz
mi/s	miles per second
min	minute
mL	milliliter
mm	millimeter
ms	millisecond
mW/cm²	milliwatts per square centimeter
n	integer index indicating number of maxima in standing wave in z direction
oz	ounce
P	fractional amount of protein in sample
P	power
P$_a$	power absorbed by a sample
P$_{avg}$	average power
P$_i$	power incident upon a sample
P$_o$	power transmitted into a sample
P$_o$	output power of microwave oven
P$_r$	power reflected from a sample
P$_t$	power transmitted through a sample
P$_v$	power absorbed per unit volume
P$_z$	power as a function of distance z into sample
PVDC	polyvinylidene chloride

pulses/s	pulses per second
r	ratio of power measured with 500 mL load to that measured with 1 L load
R	susceptor film resistivity (ohms/square)
R	process throughput (weight per time)
R_1, R_2	resistor designation
rf	radio frequency
R.M.S.	root mean square
s	seconds
SI	Scientific Internationale (International Scientific)
t	time
t_{on}	oven on time
t_{off}	oven off time
W	watt
W	weight of sample
W/m K)	watts per meter kelvin
V	potential difference (voltage)
V	volume
V_c	capacitor voltage
V_m	magnetron voltage
V_t	transformer voltage
V/m	volts per meter
V/cm	volts per centimeter
V/in	volts per inch
WR-	designation for waveguide type
Z_o	impedance or resistivity of free space (377 ohms)

Miscellaneous Symbols

°C	degrees Celcius (centigrade)
°F	degrees Fahrenheit

Greek Symbols

α	angle of incident wave
α	attenuation constant for plane wave

α	thermal diffusivity
β	angle of refracted wave
ΔT	temperature rise
Δt	elapsed time
ϵ	complex permittivity or dielectric constant
ϵ'	permittivity or dielectric constant
ϵ''	dielectric loss factor
ϵ''_σ	loss factor due to conductivity
ϵ''_d	loss factor due to dipolar effect
ϵ''_t	total loss factor from conductivity and dipolar interaction
ϵ_0	permittivity of free space
η	index of refraction
η	viscosity
κ	permittivity (infrequently used)
κ	thermal conductivity
λ	wavelength
λ_m	wavelength in a medium
λ_0	wavelength in free space
μ	complex permeability
μ'	magnetic permeability
μ''	magnetic loss factor
μL	microliter
Ω	resistance
Ω/sq	ohms per square
π	3.1415927
ρ	density
σ	electrical conductivity

MICROWAVE COOKING AND PROCESSING

Engineering Fundamentals for the Food Scientist

1

Introduction to Microwaves

An understanding of the entire field of microwave cooking, heating, and processing requires a good physical feel for what microwaves are and how they behave. Workers who wish to apply microwave technology only as an adjunct to their own specialty are not interested in a detailed mathematical explanation. Therefore, a simple but accurate physical explanation of what microwaves are and how they interact with matter is presented in this chapter.

FORCE FIELDS

Gravitational Force Field

At present, there is agreement that five fundamental forces exist in nature. Three of these we observe daily; the other two are interatomic forces and are not of interest for microwave applications. The most commonly observed force of the three is gravity. Gravity represents an attractive force between two different masses and is exhibited by the attractive force of the earth for all objects on the earth's surface. This force retains all bodies on the surface of the earth and returns virtually all bodies projected upward back to the earth's surface.*

We can thus visualize that there exists between two masses an invisible force field whose strength is directly proportional to the size or magnitude of the masses involved. An example of this size effect is the difference be-

---

*Bodies with high enough initial velocity may escape the earth's gravitational force altogether and continue into outer space, or may be captured in an orbit that circles the earth, such as a satellite.

tween the force exerted by a dropped golf ball compared with the force of a bowling ball dropped from the same height.

Magnetic Force Field

A second common force is magnetism. The earth acts as a very large permanent magnet, with north and south poles (Figure 1-1). Emanating from the north pole and terminating on the south pole are invisible lines of magnetic field that are defined as a magnetic force field. This force field can exert a force or a torque (twist) on a magnetic needle, and is the basis of our common magnetic compass. The strength of the force exerted on a magnetic compass needle, is proportional to the strength of the magnetic field and to the magnitude of the magnetic poles of the needle.

Electric Force Field

A third force field is the electric field. Under certain circumstances in some crystalline materials (semiconductors), and in virtually all metals, the negatively charged electrons of the atom orbiting the positively charged nucleus become free and are able to move throughout the material in a manner very much similar to a gas. In many materials other than semiconductors and met-

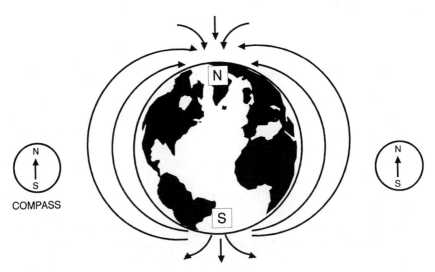

FIGURE 1-1.   The Earth's Magnetic Field   (Courtesy International Microwave Power Institute).

als, electrons can be separated physically from their positive nucleus by mechanical motion, such as rubbing or turbulence. When such a charge separation occurs, an electric force field is created between the separated charges. The strength of this field depends on the magnitude of the charges and the distance between them. Similar to the behavior of gravitational and magnetic fields, an electric field exerts a force on a charged particle, the magnitude of which is proportional to the electric field and the magnitude of the charge exposed to the field.

An example of the electric field phenomenon occurs in thunderclouds where the physical turbulence of the air causes very large scale separation of positive and negative charges. When this separation occurs, the electric field between the charges can become so high that the force it exerts upon other charged air particles in the vicinity accelerates them to a high velocity. When these high-velocity air particles strike adjacent particles, they physically break loose the electrons of their atoms, thus creating more charged particles. This acceleration phenomenon is repeated many times over, causing an avalanche effect that results in a massive flow of charged particles known as an electrical current. In this form, when current flows in air it is known as a lightning discharge and, as will be seen, is equivalent to the arcing that occurs inside a microwave oven when it is improperly used.

A technique exists, other than the mechanical motion previously described, for producing this charge separation. By complex chemical means, charges can be separated within certain chemical compounds in conjunction with specific types of metallic electrodes. Such devices are called cells or, more popularly, batteries. Batteries can produce electric fields within metallic conductors or wires and have the ability to move electrons or electric charges within the wires. This electron motion in the wires is the current flow that produces the energy for lighting, heating, and a myriad of other useful applications.

To define some of the terms utilized, consider the example of a battery connected by wires to two large parallel plates and separated by a distance d, as in Figure 1-2a. A measure of how much charge separation has occurred in a battery is given by its potential difference V (often called voltage) with units given in volts (V). Since the wires and the metallic plates are conductors of electricity, the separated charges can flow from the battery to the plates but not across the air gap. Inside the air gap, between the parallel plates, an electric field E is created. The value of this field is given by the ratio of the potential difference to the distance between the plates, with units commonly expressed as volts per meter (V/m):

$$E \text{ (V/m)} = \frac{V \text{ (V)}}{d \text{ (m)}}$$

$$(1-1)$$

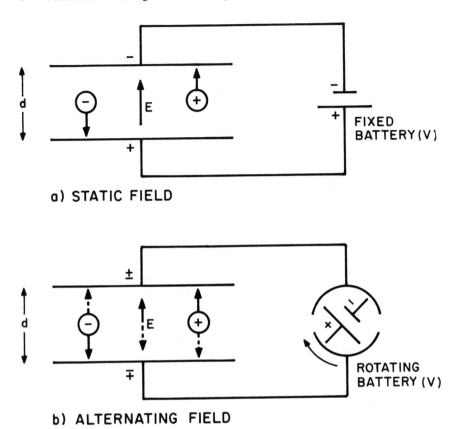

a) STATIC FIELD

b) ALTERNATING FIELD

FIGURE 1-2.   Generation of Electric Field: (a) Static Field; (b) Alternating Field.

Other units of distance are sometimes used, giving rise to electric field units of volts/centimeter (V/cm) and volts/inch (V/in). The electric field is a vector quantity, which means it has direction as well as magnitude. The direction is defined as that direction in which a positively charged particle is accelerated when exposed to the field. Negatively charged particles, such as electrons, are accelerated in a direction opposite to the direction of the electric field. The strength of the electric field increases as the potential increases or the separation between the plates decreases.

The electric field is constant or static as long as neither the potential nor distance between plates vary. To describe a time-varying electric field, on the other hand, consider the battery of Figure 1-2a located inside a cylindrical cage and allowed to rotate as shown in Figure 1-2b. The rotating battery makes sliding contact with two fixed semicircular electrodes connected by wires to the fixed plates. As the battery is rotated, first the top plate is connected to the negative electrode of the battery, and the positive electrode is connected to the bottom plate. After a quarter turn, the connections are reversed, with the bottom plate becoming negative. The electric field in the first case is directed upward and then reverses to a downward direction. Thus, by definition, a charged particle between the plates is alternately accelerated back and forth in the direction of the alternating field. The magnitude of the electric field as a function of rotation, and thus of time, is graphed in Figure 1–3*a*. The form of the graph is called a square wave. The rate at which the field alternates is determined by the speed of rotation. This rate is defined as the frequency of oscillation, *f,* and is given by the number of complete oscillations per time (usually designated in seconds^{-1}) (s^{-1}). The time to make one full oscillation is defined as the period of oscillation τ, usually designated in seconds (s). Note that the period is the reciprocal of the frequency: τ (s) $= 1/f$ (s^{-1}).

Another example of an alternating or oscillating electric field can be constructed by disconnecting the wires going from the plates to the battery cage and connecting them by means of a plug to a standard wall socket. The 120 V from the socket is alternating but not in the discrete, square-wave fashion of the battery cage. Due to the way electric power is generated, the potential *V* goes through a complete cycle beginning at zero, increasing to a maximum in one direction, returning to zero, increasing to a maximum in the opposite direction, and again returning to zero. The electric field between the plates thus alternates in a similar fashion. The magnitude of the field is graphed in Figure 1–3*b*. The form of this variation is called a sine wave and is defined explicitly by trigonometry. The frequency of a sinusoidal variation, or the number of oscillations per second of a sine wave, is defined as the electrical unit hertz (Hz).*

The alternating frequency of the power generated in the United States, for example, is 60 Hz; in Europe it is generally 50 Hz. If the frequency of the electrical generator were increased to 550,000 Hz, an electric field equivalent to the oscillation frequency of a radio station at the lower end of the amplitude-modulated (AM) radio band would be created. Further increasing the generator frequency, say to 97,100,000 Hz, produces a signal equivalent to a frequency-modulated (FM) radio station. Increasing the frequency still higher to 2.45 billion Hz, we would have the equivalent of a microwave generator in a home or commercial microwave oven.

---

*F. W. Hertz, a German physicist (1857–1894), discovered electromagnetic waves around 1886. Sinusoidal frequency, previously measured in cycles per second, was named hertz in his honor in 1960 with the adoption of an international system of scientific units (designated SI).

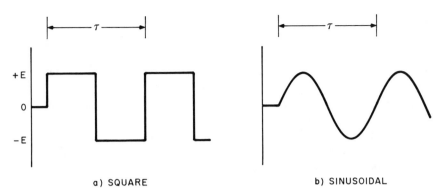

a) SQUARE   b) SINUSOIDAL

FIGURE 1-3.   Electric Field Waveforms: (a) Square; (b) Sinusoidal.

Different prefixes are commonly used to designate different magnitudes of units. Common prefixes used in microwave technology are listed in Appendix 1, along with their application to the hertz.

The AM radio oscillation frequency could be designated 550 kilohertz (kHz) or 0.55 megahertz (MHz). The microwave oven frequency of 2450 MHz could equally be designated 2.45 gigahertz (GHz). Both MHz and GHz are commonly used in the microwave oven industry.

INTERACTION OF MICROWAVES WITH MATERIALS

Ionic and Dipolar Interaction

Ionic Interaction

If an electric generator, oscillating at 2.45 GHz, is connected to the parallel plates of Figure 1-2 it will create an electric field between them that alternates one full cycle 2.45 billion times per second. In a microwave oven or in an industrial microwave system, an electronic vacuum tube called a magnetron (Chapter 2) produces the alternating electric field inside the oven cavity. This very high frequency oscillating field is called a microwave electric field, as explained in the following sections. The microwave electric field in the cavity acts in the same fashion as the electric field between the plates of Figure 1-2b. In the enclosed cavity of a microwave oven, the field is not in one single direction, but extends in the three perpendicular oven directions: top to bottom, side to side, and front to back.

Thus, any charged particle found in the oven, or in any material placed in the oven, experiences a force alternating in the three orthogonal directions at 2.45 billion times per second. The net force due to the three fields will be in some arbitrary direction in space, depending on the amplitude of the three individual forces. The net force will first accelerate the particle in one direction and then in the opposite, with particles of opposite charges being accelerated in opposite directions. If the accelerating particle collides with an adjacent particle, it will impart kinetic energy to it and set it into more agitated motion than it previously had. Since heat is defined as the agitation of particles or molecules, the particle's temperature is increased. As the second, more agitated, particle interacts with its neighbors, it transfers agitation or heat with them, until all neighboring particles have had their temperature increased. By this mechanism, energy from the oscillating microwave electric field in the microwave oven cavity is imparted to charged particles in terms of physical agitation. This increased energy of agitation, or heat, is then transferred to other parts of the material. This energy transfer from microwave field to agitated particles is the mechanism of microwave heating.

Most food products contain water with varying amounts of dissolved salts such as sodium, potassium, and calcium chlorides. When these salts dissolve, the molecule ionizes or separates into two charged particles or ions. The sodium, potassium, and calcium atoms become positively charged particles called *cat*ions, since they are attracted to a cathode or negative plate. The chlorine atoms become negatively charged particles called *an*ions, since they are attracted to an *an*ode or positively charged plate.

Food products or any material containing such charged ions are able to interact with any electric field, including the electric field in a microwave cavity. The energy transfer to the ions and then to neighboring atoms or molecules is one mechanism of microwave heating. The interaction with sodium and chlorine ions is illustrated in Figure 1-4a.

Dipolar Interaction

The water molecules prevalent in most foods and many materials are made up of two hydrogen atoms and an oxygen atom. The structure of the molecule is in the form of a V (Figure 1-4b), with the two hydrogen atoms attached to the oxygen atom making an angle of approximately 105°. The hydrogen atoms each consist of a positive charge, and the oxygen atom consists of two negative charges. As shown, the charges are physically separated and in this form are called a dipole (two poles). Molecules with such separated charges are called polar molecules. This dipole is the equivalent of an electric compass needle and acts in an electric field in the same way that a magnetic compass needle acts in a magnetic field. If these water molecules are placed in a region of an oscillating electric field, they will experience a torque or rotational force attempting to orient them in the direction of the field.

Before the microwave electric field is applied, all the water molecules in the food are thermally agitated in a random fashion corresponding to the initial temperature of the sample to be heated. When the field is applied, the molecules all attempt to orient themselves in the initial field direction.

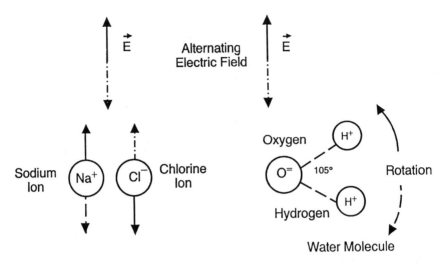

FIGURE 1-4.   Microwave Interactions (a) Ionic; (b) Dipolar   (Courtesy International Microwave Power Institute).

As they do, they collide randomly with their neighbors. When the field reverses, they attempt to reverse direction and further collisions occur. These collisions add to the background thermal agitation, which we then perceive as an increase in temperature. Energy is thus extracted from the oscillating electric field by the dipoles and is transferred to other molecules by the collisions. Molecules in solid form, such as ice, are locked into specific locations by the crystal structure and cannot rotate enough to collide with their neighbors. In a gas or vapor form, there are not enough neighboring molecules to collide with to extract appreciable power.

Microwave interaction with polar molecules such as water is the second form of microwave interaction. For foods it is the predominant microwave interaction, except in highly salted foods such as ham.

Microwaves are *not* heat. Microwave fields are a form of energy, and microwaves are converted to heat by their interaction with charged particles and polar molecules; their agitation is defined as heat.

PROPAGATION OF ELECTROMAGNETIC WAVES

Our discussion has centered only on the electric field generated by separation of electric charge or by a microwave magnetron. Several other concepts must be presented to complete the description of microwave fields.

Earlier discussions of electric field frequencies mentioned that the fields generated by radio stations and microwave ovens were simply electric fields oscillating at frequencies higher than 60-Hz power lines. One unique well-known feature of radio stations is that radio signals propagate from the generator or transmitter and are received on radios located at great distances from the generator. Thus, the electric fields from the generator somehow move or propagate through space and carry with them enough energy to operate a radio receiver.

To understand the mechanism of propagation, consider again the rotating battery of Figure 1–2b. In this case one electrode is connected to the earth, and the second is connected to the bottom of a metallic tower that is insulated from the earth. The rotating battery represents the radio transmitter; the tower represents the station antenna, as is illustrated in Figure 1–5. The Citizens' Band radio and antenna installed in an automobile or the microwave magnetron (Chapter 2) installed in a microwave oven could be represented in the same fashion.

When the negative terminal of the battery in Figure 1–5a is connected to the antenna, an electric field will extend from the earth in the direction shown. When the battery is rotated 90°, it will not be connected and the

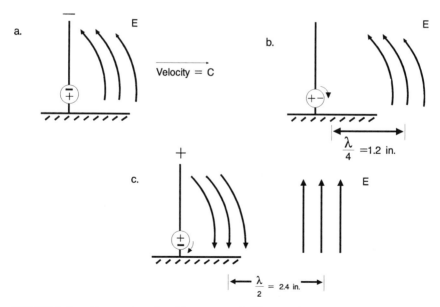

FIGURE 1–5.   Propagation of an Electromagnetic Wave   (Courtesy International Microwave Power Institute).

electric field will fall to zero (Figure 1-5b). When the battery is rotated 180°, the electric field is reversed as in Figure 1-5c.

Consider the instant when the negative terminal of the battery is first connected to the antenna. At first thought, one might imagine that the electric field adjacent to the antenna, as well as everywhere else in space, is created simultaneously upon connection of the battery. In fact, the instantaneous existence of the electric field everywhere in space cannot occur since changes in the electric field propagate with a finite speed as predicted by Maxwell (1873). This speed is the speed of light, c, (186,000 mi/s $= 3 \times 10^8$ m/s $= 3 \times 10^{10}$ cm/s) and, as postulated by Einstein and confirmed in many experiments, is the maximum speed at which energy can travel. Thus, by the time the generator has rotated 90° or one-quarter cycle, the electric field has propagated at the speed of light a distance away from the antenna. As the generator continues to rotate, alternating regions of upward, zero, and downward electric fields will be generated and propagate outward in all directions from the antenna.*

The time to complete one rotation or cycle of the battery (generator), is defined as the period τ and is $\tau = 1/f$. The distance that an electric field maximum moves away from the antenna in this time is defined as the wavelength, λ. Since distance is velocity multiplied by time:

$$\lambda \text{ (m)} = \frac{c \text{ (m/s)}}{f \text{ (Hz)}}$$

(1-2a)

A useful approximation for microwave work in English units is

$$\lambda \text{ (in)} = \frac{11.8 \text{ (in/nsec)}}{f \text{ (GHz)}}$$

(1-2b)

The wavelength at the microwave oven frequency of 2.45 GHz is thus 0.128 m, 12.8 cm, or 4.82 in.

In actuality, radio and television transmitters, as well as magnetrons, do not produce square-wave bursts of electric field but fields with sinusoidal variation, as illustrated in Figure 1-3b. If the point on the graph where the amplitude starts at zero represents antenna position, then the graph represents a snapshot of the amplitude of the electric field in space. Zero

---

*The previous description of propagation is not strictly correct in that the propagation of the changing electric charge moving to the top of the antenna is not taken into account. The description is sufficient, however, to conceptionalize the phenomenon.

amplitude at the antenna corresponds to the battery position of Figure 1–5*b*. One-quarter wavelength away, the field has maximum amplitude; one-half wavelength away the amplitude is zero, and so on.

Similarly, as a function of time, a charged particle, or a radio receiver located a distance away from the antenna experiences a force due to the electric field alternating at the frequency of the transmitter.* The force is due to the propagating regions of field passing by at the speed of light.

The Magnetic Field

The radio-frequency generator produces an electric field by alternately placing a positive charge and a negative charge on the antenna. This alternating charge is caused by moving electrons in the metallic structure of the antenna. From basic physics, a moving charge produces a magnetic field that encircles the charge-containing conductor. For the antenna illustrated, when the voltage on the antenna is increasing in the negative direction, the adjacent electric field increases in the upward direction (Figure 1–5*a*). Electrons flowing up the antenna create magnetic field loops that encircle the antenna. At a point to the right of the antenna the magnetic field is oriented perpendicular to the page and in a direction out of the page. When the electric field is zero, the magnetic field is zero. As the electric field increases in amplitude in the downward direction, the magnetic field increases in a direction into the page. The magnetic field thus produced propagates outward along with the electric field. The two fields are orthogonal (at right angles to each other) and travel at the speed of light, *c*.

In the region near the antenna, called the near field, the actual fields are complex, having curved configurations. Away from the antenna, in the region called the far field, the curvature essentially disappears. If one stands parallel to the antenna in the far-field region, the electric field would appear vertical and the magnetic field would be orthogonal or horizontal. The amplitude of both fields would alternate sinusoidally as shown in Figure 1–3*b*. Such a wave in the far-field region with orthogonal fields is defined as a plane wave.

---

*The electric field in the vicinity of a radio or television receiver accelerates the electrons in its antenna. These moving electrons are a weak electrical current that the receiver amplifies. Speech (audio) and picture (video) information are contained in the varying amplitude or frequency of this current and are decoded by the receiver. Varying the amplitude of an electromagnetic signal is called amplitude modulation (AM); varying the frequency is called frequency modulation (FM).

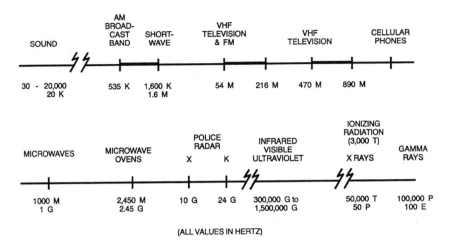

(ALL VALUES IN HERTZ)

FIGURE 1-6.   The Electromagnetic Spectrum   (Courtesy International Microwave Power Institute).

THE ELECTROMAGNETIC SPECTRUM

All waves of the type previously described are known as electromagnetic waves. Since they propagate or radiate through media and space, they are also referred to as electromagnetic radiation. All electromagnetic waves have the following common characteristics:

1. Physical behavior described by mathematical equations developed by Maxwell (1873).
2. Electric and magnetic fields orthogonal in the far-field region, with amplitudes that vary sinusoidally.
3. Propagation through empty space with the velocity of light. In other media, propagation velocity is reduced by the index of refraction of the media (Chapter 5).
4. Exhibition of characteristics of reflection, absorption, and transmission when interacting with media.
5. Relationship between frequency and wavelength is simple (cf. Equations 1–2).

It has been shown that there is basically no difference between the fields of the low-frequency radio and television waves and the higher-frequency microwave fields in the microwave oven except for the rate at which they alternate their direction. Similarly, there is little difference between the oscillating electric field of the microwave signal and that of an infrared, visible, or ultraviolet light wave, except that, again, the light signals have much

higher oscillation frequencies than do the microwaves. The one difference—the ability of the electromagnetic signal to penetrate a given material—is discussed in Chapter 5. This collection of radio, television,
microwave, and light waves is known as the electromagnetic radiation spectrum. A comparison of uses of the frequencies of this spectrum is shown
in Figure 1-6.

References
Maxwell, J. 1873. *Treatise on Electricity and Magnetism.* Oxford: Clarendon Press.

2

Understanding the Microwave Oven

Microwave heating and cooking is an outgrowth of *radar* (*ra*dio *d*etection *a*nd *r*anging) technology developed during the Second World War. During this period, vacuum tubes called magnetrons were invented and perfected that were able to generate many kilowatts of electromagnetic power at frequencies heretofore not obtainable. These frequencies, called microwaves, ranged from 1 to 30 GHz, with corresponding wavelengths from 30 to 0.3 c. The present definition for the range of microwave frequencies is 300 MHz (0.3 GHz) to 300 GHz.

Even though there are many anecdotal references to microwave technicians warming themselves in front of radar antennas, the first patent application for the application of microwave energy to heating foods was filed by Percy Spencer of Raytheon Corporation in 1945 (Spencer 1950), followed by his invention of the microwave oven in 1947 (Spencer 1949). His invention consisted of two microwave-generating magnetron tubes, a power supply to provide the appropriate voltages for the tubes' operation, a means of controlling the amount of power generated, a timer, and a means of directing the microwave power to the load to be heated, in this case a metallic box or cavity. To the present day, this general concept is used for all microwave heating and cooking apparatus, whether consumer, commercial, or industrial. An excellent review of the history of microwave cooking and heating is by Osepchuk (1984).

THE MICROWAVE HEATING SYSTEM

A simplified schematic representing a consumer or commercial microwave oven as well as an industrial processing system is shown in Figure 2-1. A cutaway view of a typical microwave oven is shown in Figure 2-2. The key

14

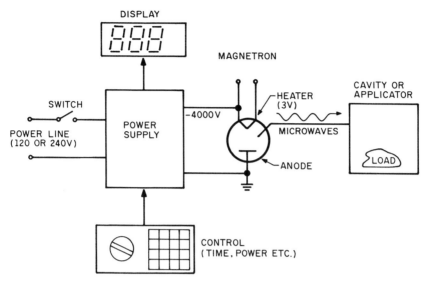

FIGURE 2-1.   Block Diagram of Microwave System.

element to the oven is the *magnetron,* which generates the microwave power (Figure 2-3). A typical consumer or commercial microwave oven magnetron requires approximately 4000 V for its operation. Industrial magnetrons require up to 7000 V and may be powered by a 440-V industrial power line. The power supply portion of the system raises the voltage from the power line (or main) to the required value by means of a transformer. Also, because the magnetron operates on a constant voltage rather than the alternating voltage supplied by the power line, a circuit of diodes and capacitors is used to change (rectify) the alternating line voltage. Due to practical mechanical design considerations, discussed later, a negative voltage is generally provided to the magnetron.

Depending on the application, the heating rate of the load must be controlled. The amount of microwave power supplied to the load determines the heating rate and is adjusted electronically within the oven by the control circuitry. Microwave systems also have a time control for adjusting the duration of microwave power application. A means for inputing power and time information into the system and a means of observing performance are also provided. In inexpensive ovens, time and power are controlled by simple timer motors and cams. Top-of-the-line ovens use microprocessor chips to perform these functions. Microprocessors also facilitate other features such as automatic cooking, delayed-start cooking, multiple-step programming, time-of-day clock, and a myriad of sensor technologies. No

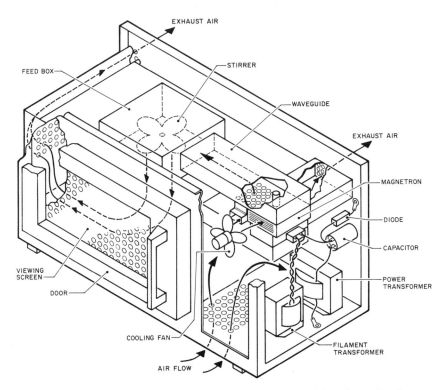

FIGURE 2-2.   The Microwave Oven   (Courtesy of Amana Refrigeration, a Raytheon Company).

matter what control technique is used, the basic oven structure and operation are the same.

Finally, when the microwaves are generated, they must be directed and applied to the load or product to be cooked or heated. In microwave ovens the microwave cavity performs this function. In industrial systems, cavities as well as special applicators are utilized.

THE MAGNETRON; THE MICROWAVE GENERATOR

The magnetron is basically a simple vacuum tube surrounded by a support frame with attached cooling fins. Emanating from the top is a 0.8-in (2-cm) long protrusion or antenna that radiates the generated microwave power in a fashion similar to a "walkie talkie" transmitter. A photograph of a 700-watt (W) magnetron used in microwave ovens is shown in Figure 2-3, and

FIGURE 2-3.   Microwave Magnetron   (Courtesy of Amana Refrigeration, a Raytheon Company).

a cutaway diagram is shown in Figure 2-4. The antenna on the top and cooling fins in the center are easily distinguishable. The small box on the bottom is called a *filter box.* The two terminals on the front of the box, as will be described, are used to connect the power supply to the magnetron filament. The box contains a pair of inductors and capacitors that filter or prevent any spurious electromagnetic emissions from escaping along the leads to the power supply.

The internal structure of the magnetron that generates the microwave power is called a *diode,* a contraction of *di* (meaning "two") and electr*ode.* The diode of a consumer magnetron consists of a cylindrical copper tube approximately 1.75 in (4.5 cm) in diameter by 1.25 in (3.2 cm) in height. The tube is capped at both ends with cylindrical copper plates so that the entire region can be evacuated to produce a vacuum.

Inside the tube and extending toward the center like spokes of a wheel are usually 12 copper plates or vanes. These vanes do not extend completely to the center, but leave an empty cavity approximately 0.375 in (0.95 cm)

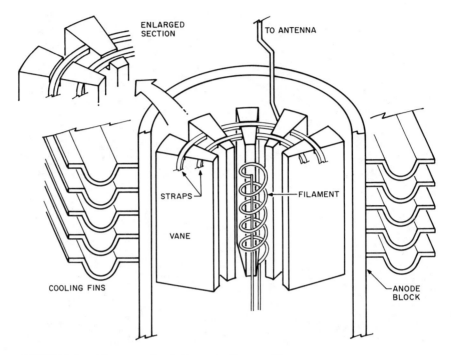

FIGURE 2-4.   Magnetron Anode Structure; Cutaway View.

in diameter in which the spiral wire filament is located. The copper cylinder and associated vanes is called the *anode*.

A simplistic explanation of magnetron operation follows and is illustrated by the cross-sectional view of the anode in Figure 2-5. When the filament is heated, electrons are "boiled" from its metallic surface and form an electron cloud in the center of the diode. When 4000 V are applied between the anode and filament (anode positive and filament negative), an electric field E is produced that accelerates the electrons radially away from their cloud toward the anode. Above and below the anode-filament structure are two ring magnets, usually ferrite. These magnets produce a magnetic field H that is parallel to the axis of the filament and anode cylinder and thus perpendicular to the electric field.

The electrons, moving outward like the spokes of a wheel, are forced into a curved path by the magnetic field. If the strength of the applied magnetic field is adjusted properly, the electrons just skim by the tip of the vanes without striking them. If the field is too weak, the electrons collide with the vanes; if too strong, they curve back and collide with the filament.

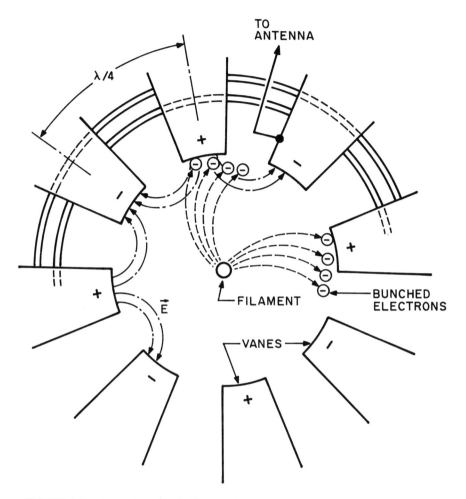

FIGURE 2-5.   Magnetron Anode Cross Section.

As a negatively charged electron illustrated by ⊖, approaches a vane, it induces an equal but positive charge in the vane. If that vane is connected electrically to another located two vanes away, it transfers its positive charge to it also. The procedure of connecting every other vane together is called *strapping,* the electrical connections being the straps. The vane located between the two positively charged vanes has an opposite negative charge induced in it. If these vanes are strapped together in a similar fashion, they will all be negatively charged. Alternate or adjacent vanes will thus be positive and negative. Since unlike charges attract and like charges repel, those electrons finding themselves adjacent to positively charged vanes will be accelerated; those adjacent to negative vanes will be retarded. The retarded electrons fall behind and meet the accelerated group forming bunches of electrons. As these bunches pass by successive vanes, they induce a positive charge in it. The vane they have just passed changes from positive to negative. As the electron bunches continue in their circle, they force any single vane to continuously alternate charge from positive to negative and back again. If the speed of these electrons is adjusted in concert with the spacing between vanes, it is possible to produce an alternating charge on a vane of 2.45 billion times per second. This alternating charge connected by a wire from any arbitrary vane to the antenna cap produces a radiating, 2.45-GHz microwave signal exactly as shown in Figure 1-5. For a more detailed discussion of the magnetron refer to Decareau and Peterson (1986).

POWER SUPPLY

As discussed later, the magnetron requires external cooling fins attached to the anode. For a simpler mechanical design, the fins are attached to the magnetron body, which is then attached to the metallic portion of the oven. This "grounding" of the anode requires a negative voltage to be applied by the power supply to the filament to accelerate the electrons toward the anode.

For consumer, and most commercial, microwave ovens a half-wave doubler power supply circuit is utilized to provide this voltage (Figure 2-6). In this type of circuit, half of the voltage, $V_t = 2000$ V, is supplied by the transformer and is illustrated in Figure 2-7a. This voltage is doubled to 4000 V by the capacitor-diode combination as follows. During the first half of the 60-Hz output from the transformer, current flows through the capacitor and diode to ground, charging the capacitor to $V_c = 2000$ V. During this half-cycle, the diode short-circuits the voltage across the magnetron so that $V_m = 0$. When the transformer voltage V_t, reverses during the second half-cycle, the diode, which only allows current to flow in one direction, can no longer conduct, thus removing the short circuit across the magnetron. The transformer voltage V_t now adds to the capacitor voltage V_c, and both appear across the magnetron as $V_m = 4000$ V. Current flows through the magnetron, microwave power is generated, the capacitor discharges, and the cycle is repeated. Note that no microwave power is generated during each first half-cycle of the line voltage when the capacitor is being charged. In actuality, microwave power is generated in 8.33-ms bursts (Figure 2-7b). Except for very special scientific applications, such as the microwave generation of some gaseous plasmas, the load being heated cannot distin-

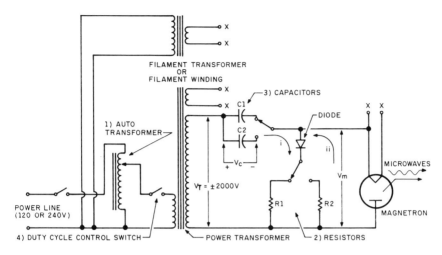

FIGURE 2-6.   Microwave Oven Power Supply Diagram.

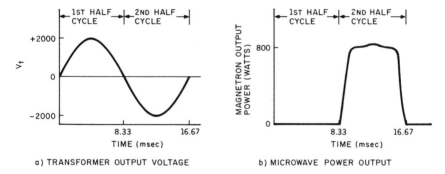

FIGURE 2-7.   Typical Waveforms: (a) Transformer Output Voltage; (b) Microwave Power Output.

guish the sequence of short microwave pulses from continuous microwave power.

POWER CONTROL

The microwave output power from a magnetron can be varied by a number of techniques. The paragraph numbers below are referenced on Figure 2-6.

1. Variable Voltage Supply.   If a means is provided to vary the line voltage supplied to the transformer, the magnetron output power can be varied. Variable or auto transformers (1) as well as electronic circuits are available for this purpose, but they are expensive and infrequently used. The sensitivity of output power to line voltage is a source of concern to the microwave oven food developer as well to as the consumer. Measured oven power as a function of line voltage is shown in Figure 2–8 for three production ovens. Note the variation of the 500-W oven (oven 2), with a 6% change in line voltage. A proposed European standard would require less than a 6% change in microwave power output for a similar change in line voltage. Oven 2 falls well outside of this specification.

   2. Resistive Control.   Placing a resistance in series with the diode limits the amount of charge, and thus the voltage, V_c that the capacitor receives. This reduction in voltage controls the microwave power as in (1) above. Power can be controlled by switching resistors (2). The technique is inefficient, however, because the resistor must dissipate the unused power as heat, so the concept has been used sparingly. Using a resistor with a value of 1 or 10 ohms (Ω) in this place in the circuit and measuring the voltage across it is an excellent means for monitoring current being supplied to the magnetron. Magnetron output power is a linear function of this current, which is represented by the voltage measured across the resistor, thus providing an excellent technique for monitoring oven power. A mechanical-

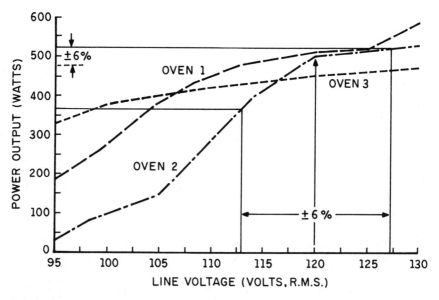

FIGURE 2-8.   Oven Power vs. Line Voltage.

movement voltmeter is preferable, because it mechanically averages the pulsed 8.33-ms power pulses.

3. Capacitive Control. By switching capacitor (3) values, we can vary V_c. This technique is difficult because the switching must take place at high voltages. It has thus seen only limited application. The sensitivity of power to capacitance illustrates a practical manufacturing problem. To hold tight oven power specifications, capacitors must be measured and selected at high cost. Tolerance variation of capacitors is a common cause of power variation within an oven model family.

4. Duty Cycle Control. Virtually all microwave ovens use duty cycle control for varying oven power. This technique simply turns the oven on and off at full power automatically. The longer the oven is turned on compared with the time it is turned off determines the amount of average power, P_{avg}, delivered to the load. The technique is illustrated in Figure 2–9a for an average power of 35%. The oven is turned on for a time $t_{on} =$

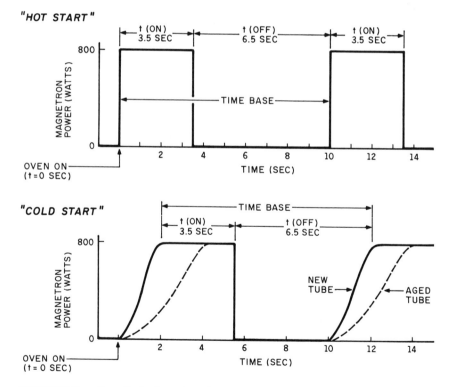

FIGURE 2-9.   Duty Cycle Control: (a) Hot Start; (b) Cold Start.

3.5 s and then turned off for a time $t_{off} = 6.5$ s. The cycle is then continuously repeated. The commonly used terms of duty cycle and time base are defined as

$$\text{Time base} = t_{on} + t_{off} \tag{2-1}$$

$$\text{Duty cycle} = \frac{t_{on}}{\text{time base}} \tag{2-2}$$

The duty cycle thus represents the fraction of time that microwave power is applied to a load. It also is commonly expressed as a percentage by multiplying by 100%. The average magnetron output power is $P_{avg} =$ duty cycle × magnetron output power. For the duty cycle of 0.35, $P_{avg} = 280$ W for an 800-W oven. Note that the 3.5-s pulse is actually made up of 3.5 s × 60 pulses/s = 210 pulses of the type shown in Figure 2–7*b*.

Hot-Start and Cold-Start Ovens

The magnetron filament must be heated to approximately 3180°F (1750°C) for it to properly emit electrons. The heating is provided by an electrical current applied to the filament by means of a small filament transformer. In early ovens, this filament transformer was a separate unit that applied 3–4 V to the filament. In Figure 2–6 the filament transformer terminals labeled x-x are connected to the magnetron filament terminals x-x. With such a circuit, the filament is energized when the oven is first turned on and continues to emit electrons during oven operation. When the duty cycle control switch (4) applies power to the magnetron, microwave power is generated essentially instantaneously, and the output of the magnetron appears as illustrated in Figure 2.9*a*.

An oven with a filament transformer is called a hot-start oven because the duty cycle control is turned on when the filament is hot. The advantage of a hot-start oven is that the time base for power control can be made extremely short. Short time bases below 2–3 s eliminate the problem of pulsing or "breathing" of foods when time bases are long. For example, a sauce being prepared at 35% power with a 10-s time base rises in a bowl for 3.5 s and then collapses for 6.5 s. This breathing is not only annoying but can be a problem for sensitive foods that cannot tolerate heating by such high power bursts. Time bases for hot-start ovens are usually a few seconds. Some ovens even operate with a time base as low as 167 ms or 10 line cycles.

As competition increased among oven manufacturers, cost reductions became common. This was shown when the 50-cent cost of the filament transformer, that sold at retail for between $2–$3, was eliminated in favor of a few extra turns of wire on the power transformer, which cost only a few cents. The filament winding alternative is also illustrated in Figure 2-6, with connections to the filament shown as x-x. With this circuit, the filament is turned on and off along with the duty cycle control. Since the filament requires approximately 1.8–2 s to become hot enough to emit electrons, microwave power is generated only after this delay. The power output of such an oven is illustrated in Figure 2-9b. Virtually all consumer ovens on the market use the filament winding or "cold-start" circuit.

There is one major and two minor disadvantages of the cold-start circuit over the hot-start:

a. Long Time Base.   Because of the 2-s warmup time, it is difficult to provide high duty cycle or high average power with a short time base. As an example, with a 2-s warmup time, an oven with a time base of 4-s could not have an average power control over 50%; i.e., $t_{off} = 2$ s during warmup plus $t_{on} = 2$ s. Typically, cold-start ovens have time bases of about 12–15 s, thus having variable power control up to 83% to 87%. Ovens on the market have been measured to have time bases between 1 and 32 s (Gerling 1984). Breathing with variable power control cold-start ovens is a serious disadvantage.

b. Aging.   The heated filaments of magnetrons deteriorate with age. Their electron-emitting efficiency declines very slowly at first and after 7 to 15 years declines rapidly until the tube fails. As the filament efficiency falls, the output power of the magnetron decreases (Figure 2-10). The consumer

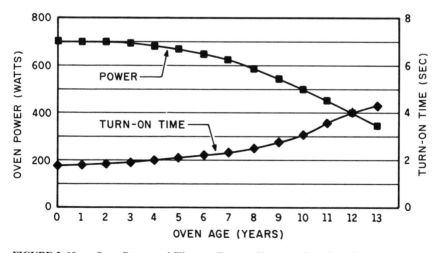

FIGURE 2-10.   Oven Power and Filament Turn-on Time as a Function of Age.

compensates for this power-aging effect by unconsciously increasing the preparation time setting. Near the end of oven life, consumers may be aware that their oven "cooks slower than it used to."

At the same time that power decreases, warmup time increases from 1.8 s to as much as 5 s (Figure 2–10). Increasing warmup time with age produces a severe problem with variable power settings. Consider the cold-start example of Figure 2–9b, where a 0.35 (35%) duty cycle is illustrated. As the tube ages from a 2-s to a 4-s warmup time, note that the actual time the magnetron is producing power drops from 3.5 to 1.5 s. A defrost setting of 35% will therefore barely work near the end of oven life. The only practical solution to this problem is for the oven operator to set the power control to higher values as the oven ages.

c. Moding.   In cold-start ovens, the high voltage and filament voltage are applied simultaneously to magnetrons. During the warmup time, stress is placed on the filament by the applied high voltage, causing the magnetron to jump to an abnormal mode of operation. This abnormal situation shortens magnetron life. Most magnetrons have incorporated design changes that alleviate this problem. In today's microwave ovens, moding usually occurs only near the end of magnetron life or with defective tubes.

5. Switching or High-Frequency Power Control.   A number of manufacturers have introduced from time to time a radically new type of power supply that allows essentially continuous power control, at least over a major portion of the power range. These power supplies are sometimes also called switch-mode or solid state supplies. These supplies change the 60-Hz line voltage to a much higher frequency, around 25–35 kHz. The high-frequency voltage is produced electronically by switching semiconductors. The advantage of this technique is that the transformers needed to step up the voltage to operate the magnetron are reduced in size and weight by an order of magnitude. Because the voltage is switched electronically, it is simple to control the width of the 35-kHz pulses. Thus, duty cycle control can be performed with a time base of 1/35,000 s. Duty cycle control with such a short time base is indistinguishable from continuous control. The high cost of switching semiconductors and ancillary electronics may be offset by lower transformer costs as well as considerably lower shipping costs, where shipping is a consideration. Some switching power supplies may weigh almost 12 lb less than normal.

THE OVEN STRUCTURE: DOOR, CHOKE, CAVITY, STIRRER, AND FEED

If the antenna of the magnetron were allowed to radiate directly into space, the generated power could not be effectively utilized. To direct the application of power to the desired load, the antenna is usually inserted into a rectangular tube called a *waveguide*. The waveguide channels the micro-wave radiation either directly into the oven cavity or first into a feedbox with stirrer and then into the cavity where it heats the food (Figure 2–2). Stirrers are used to smooth out the hot and cold spots in the oven and will be discussed subsequently.

Door and Choke

Access of food to the oven cavity is through a carefully engineered door structure. Integral to the door, but not visible, is a microwave circuit called a choke. This choke prevents microwave radiation from leaking through the minute gaps that can exist between the door and the cavity face plate. Although choke operation is complex, it can be likened to an electrical light switch. When the switch is closed, power can flow through the two wires to the light. If the switch is opened, a gap or open circuit is created in one of the conductors, preventing current flow. The door choke provides such a gap, preventing excessive microwave leakage (cf. Chapter 9 for a discussion of regulations and safety of leakage).

The Cavity

One of the major problems associated with microwave cooking is the production of hot and cold spots within the food product. A portion of this problem is due to the oven and is caused by the high and low values of electric field that are intrinsic to the cavity structure. The geometry and dielectric properties of the food also affect heating uniformity, and are discussed in Chapter 6. Unfortunately, most uneven heating results from fundamental laws of physics and cannot be completely eliminated.

The electric field variations in a cavity occur for the following reason. Microwaves entering the cavity are reflected from the metal walls in exactly the way light is reflected from a silvered mirror. The reflection occurs because the electric field parallel to the metal walls is "short-circuited" by the conducting metal and is forced to a value of zero at the wall. Since there is microwave energy inside the cavity, the electric field is other than zero internal to the cavity. The amplitudes of the microwaves bouncing around in the cavity add and subtract from each other, producing regions of high and low electric fields, (i.e., hot and cold spots). The distributions of these stationary patterns of electric field in the three dimensions of the cavity are called *standing waves*.

Another way of understanding the electric field pattern in a microwave oven is to visualize a rubber band stretched between two sides of an oven cavity. The rubber band, attached to the cavity walls, is unable to vibrate at either end, being held to zero amplitude. If the band is plucked in the center, it will vibrate with a maximum amplitude at the center. This configuration is called the fundamental mode of vibration. If the band is plucked one third of the way across the cavity, it will vibrate with maximum amplitude at one third and two thirds of the distance. The amplitudes are equal, but in opposite directions at the two maxima. In this, the second harmonic mode, there are three positions of zero amplitude: the two walls and the cavity center. The plucking illustration can be carried further, with vibrations having four, five, or more, maxima. There will always be one more zero associated with a vibration mode than there are maxima, if wall zeros are counted. If the band is plucked at an arbitrary position, many modes of vibration will be generated, producing an arbitrary vibration pattern. Under this condition there will probably not be any region of exactly zero amplitude except at the walls.

The electric field amplitude distribution in a cavity can be represented in exactly the same way as the rubber band vibrational amplitude distribution. Modes with one, two, three, or more, electric field maxima and corresponding minima may occur. In a microwave cavity, unlike the rubber band example, only modes with up to four to six maxima can exist, depending on cavity dimensions. The manner in which the microwave power is introduced into the cavity determines which modes are generated or "excited." The final amplitude distribution, and thus the hot and cold spots, depend on this excitation.

In a cavity, modes, and thus hot and cold spots, exist in all three dimensions. Thus, heating uniformity not only depends on placement of the food in the oven but also on height above the shelf. Figure 2–11 illustrates three modes in orthogonal directions, with one, two and three maxima, for an oven cavity.

The frequency of the modes in a cavity is related to its dimensions by

$$f\,(\text{GHz}) = c \left\{ \left(\frac{l}{2A} \right)^2 + \left(\frac{m}{2B} \right)^2 + \left(\frac{n}{2C} \right)^2 \right\}^{1/2}$$

(2-3)

Here f is the frequency of the mode; l, m, n, are the number of maxima in a mode in the x, y, and z dimensions of the cavity; A, B, and C are the x, y, and z dimensions of the cavity in meters or inches (Figure 2–11); and c is the velocity of light: 0.30 m/ns or 11.8 in/ns. To calculate the modes typically excited in a 2.45-GHz oven cavity, we choose a frequency range from 2.4 to 2.5 GHz. A computer program can easily scan all modes up to $l = m = n = 10$ for a given cavity dimension and select those that fall within the above range. Typical microwave ovens do not have modes with l, m, or n greater than 7.

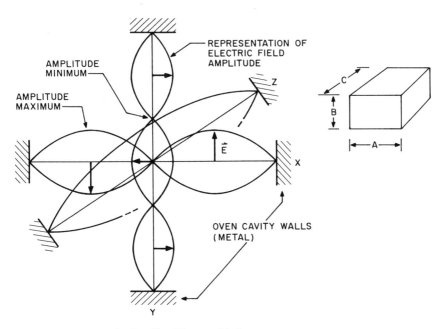

FIGURE 2-11.   Cavity Standing Waves or Modes.

As an example, calculated frequencies for cavity dimensions *A, B,* and *C* equal to 14, 10, and 12 in (35.6, 25.4, and 30.5 cm), respectively, are given in Table 2-1. (Note that modes with any two of the indices l, m, or n equal to zero cannot exist.)

Stirrer and Feed

As indicated, spatial variation of electric field amplitudes, or hot spots, of the cavity modes throughout an oven are one contributor to the uneven heating characteristic of microwaved foods. To reduce the effect of these hot spots a device called a *stirrer* is commonly used. The microwave energy from the magnetron is typically fed from the magnetron to the cavity via a waveguide to a feedbox containing a stirrer. The stirrer is nothing more than a metallic fan blade that rotates within the feedbox and is rotated by a small motor or by the air supplied by the magnetron cooling blower. After the stirrer is rotated, the air is usually directed into the cavity to purge it of any moisture generated by the cooking process. The air is eventually vented to the outside (Figure 2-2).

The stirrer can be thought of crudely as "reflecting" the microwaves

TABLE 2-1   Mode Frequencies of Microwave Oven Cavity

Mode Index Number			Mode Resonance Frequency
l	m	n	f (GHz)
0	0	5	2.461*
0	4	1	2.413
1	0	5	2.497
1	4	1	2.450
2	2	4	2.446
2	3	3	2.456
3	1	4	2.414
4	3	0	2.447
4	3	1	2.496
5	1	2	2.401
5	2	0	2.417
5	2	1	2.467

Frequency range: 2.4–2.5 GHz.
Cavity dimensions: A (width) = 14 in (35.6 cm).
             B (height) = 10 in (25.4 cm).
             C (depth) = 12 in (30.5 cm).
*Mode cannot exist.

around within the cavity, evening out the hot and cold spots. More correctly, the stirrer varies how each of the modes is excited and thus varies each mode's amplitude. The sum of the amplitudes of all modes provides a time-varying electric field over the product, thus effectively averaging out the hot and cold spots. Ovens with dual waveguides, feedboxes, and stirrers have been marketed (Litton 1978) with excellent cooking performance. Costs prevented these models from remaining competitive as off-shore competition increased.

To further help with nonuniform heating, most ovens have a microwave transparent shelf that is elevated from the metallic bottom of the cavity. This shelf is usually glass or ceramic and is either permanently installed or in the form of a removable tray. The shelf allows microwave power to reflect off the metal bottom of the cavity and impinge on the underside of the food. This technique eliminated poor bottom heating.

Most ovens with stirrers feed the cavity from the top. One manufacturer, however, has been very successful in designing a product line of bottom-feed ovens with stirrers mounted below the shelf (Quasar 1991). In these units, power enters the oven through this microwave transparent shelf with surprisingly good heating uniformity.

In ovens with a turntable, the microwaves are fed directly into the cavity from the top or side without the use of a stirrer. The turntable rotates the food product through the fixed hot and cold spots, providing a spatial averaging effect. Even though cooking performance is improved, few, if any, ovens have both a stirrer and a turntable.

References

Decareau, R. and Peterson, R. 1986. *Microwave Processing and Engineering.* Chichester, England: Ellis Horwood.

Gerling, J. 1984. *Correlation Data and Analysis.* Report 84-001, Gerling Laboratories, Modesto, CA.

Litton 1978. *Meal-in-One Microwave Cookbook.* Litton Microwave Cooking Products, Sioux Falls, SD (presently doing business as Menumaster).

Osepchuk, J. 1984. A history of microwave applications. *IEEE Transactions on Microwave Theory and Techniques* **32**(9):1200–1224.

Quasar 1991. *Quasar Microwave, Microwave/Convection Ovens.* Brochure #49900. Quasar, Division of Matsushita, Elk Grove Village, IL.

Spencer, P. 1949. U.S. Patent 2,480,679.

Spencer, P. 1950. U.S. Patent 2,495,429.

3

Sensors

A prime goal in the preparation of food is to obtain a finished product that has exactly the organoleptic qualities desired. Conventional cooking usually relies on the skill of the cook as well as a series of guidelines, such as cooking a given number of minutes per pound at a given temperature. Microwave preparation, being a newer technique, requires the learning of a new skill. To facilitate microwave cooking, oven manufacturers have incorporated in many of their models different techniques to indicate product doneness. These techniques include built-in look-up tables to determine cooking time if the product and weight are known, as well as various types of sensing devices. Unfortunately these techniques have yet to gain wide popularity, partly for technical reasons and partly for lack of advertising and well-trained sales personnel. Ultimately the failure of sensor schemes is due to the inability to provide sufficient information to the oven controller to accurately predict doneness.

LOOK-UP ALGORITHMS

Extrapolating from conventional cooking, if the category and weight of a food product are known, a time per unit weight (e.g., minutes per pound) constant for that food can be determined. This information can be stored in the microwave oven controller and "looked up" and used to set the preparation time when food category and information indicating weight are entered. Weight-indicating information may be in the form of pounds, cups, or other units, such as number of plates to be heated or bunches of broccoli to be cooked. Litton Microwave Cooking pioneered this technique with its Autocook oven (Buck 1984), and other manufacturers have followed with a wide variety of adaptations. To date the look-up algorithm technique has been popular and successful.

TEMPERATURE SENSORS

As with conventional preparation, the internal temperature of a product is an excellent indicator of its doneness. Thermometers designed specifically for use in a microwave oven are readily available on the retail market. Also, since 1976 (Chen 1976), most microwave oven manufacturers have at least one model available with a built-in temperature probe. These probes, are inserted into the food being cooked, and usually indicate the internal temperature on a digital display. They also may be preset to turn the oven off after the food has reached a preset temperature. As described in Chapter 6, cooking may continue after microwave power is turned off. Thus for microwave preparation of foods, especially meats, food must be removed from the oven at a lower temperature than is the case for conventionally prepared foods.

Built-in probes have several disadvantages. First, due to pricing pressure on microwave ovens in the marketplace, the temperature probes may not be accurate, except for models at the high price end of the product line. Errors of up to ± 8°F (3°C) may occur, with the potential for jeopardizing the quality of an expensive cut of meat.

If the probe is used for measuring the temperature in thin food products, another error may be introduced. Because the probe is usually a thin, pointed, metal rod, it may act as an antenna, picking up and concentrating microwave energy in the vicinity of the probe. This phenomenon may occur in thin foods, where microwave energy is able to penetrate to the vicinity of the probe. The concentrated energy produces localized heating and provides a temperature indication that may be considerably higher than the body of the food itself. As a result, probe placement is a major problem even for educated consumers.

Since the probe is connected by a cable to a jack in the oven wall, it is possible that the cable might become caught in the door when closed, thus causing microwave leakage. Therefore, Underwriters Laboratories (1983) requires oven manufacturers to prevent this occurrence. Door entrapment is now sometimes prevented by spacing beads along the probe cable. Probes obviously should not be left in an empty oven, whether connected or disconnected from the jack. Metal-to-metal contact in the cavity may cause arcing if the oven is inadvertently turned on with the probe inside.

Temperature probes in microwave ovens can offer a reasonable measure of food doneness. They thus make an excellent marketing tool for high-end models. Strangely, it has been found that less than 18% of the consumers who buy ovens with probes use them (Anonymous 1991). Reasons for their unpopularity have been speculated to be essentially the disadvantages already cited (Sadlack 1985).

HUMIDITY SENSING

With virtually all heating and cooking procedures, moisture evaporation occurs. It is thus technically viable to expect a correlation between food doneness and moisture lost. In the late 1970s many manufacturers explored the possibility of measuring moisture lost by cumulatively summing the absolute humidity as measured in the exit ventilation air stream in the oven. Efforts failed due to the lack of adequate accuracy of humidity sensors when utilized in the dirty environment found in a microwave oven.

Matsushita partially succeeded by requiring most foods to be covered and using the humidity sensor to detect the gross change in humidity when the pressure in the vessel became high enough to release a "puff of steam" into the cavity (Kobayashi and Kanazawa 1982). An algorithm built into the oven controller extrapolates the amount of remaining preparation time required, depending on the category of food preprogrammed into the oven. Aside from the requirement that the food be covered, sensor contamination was also a problem. The solution of requiring an internal heater to clean the probe prior to each use dictated an unacceptable delay up to 1 min before cooking could commence.

Several manufacturers attempted to use an inexpensive gas sensor, readily available in Japan, to sense the escaping gas or water vapor in a fashion similar to the Matsushita patent (Tanabe and Nakagawa 1982). This approach, however, has met little success due to humidity measurement inaccuracies.

WEIGHT SENSING

If the moisture lost from a food can be correlated to doneness, then the weight lost from the food itself should also be correlatable (Figure 3-1). Amana introduced a weight-sensing oven using this technique around 1985 (Robeson 1985). The oven exhibited good performance but met with little success, perhaps due to cost-to-value considerations. The principle, however, remains sound.

Weighing technology has been used with success in analytical instruments used to measure moisture content of materials. The use of microwaves for drying samples during weighing provides a much more rapid measurement than can be obtained by conventional techniques (Borowski and Fritz 1990).

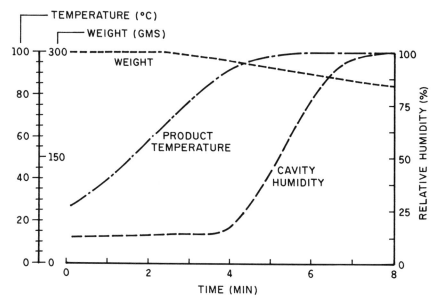

FIGURE 3-1.   Product Temperature, Weight, and Cavity Humidity as a Function of Time.

INFRARED AND VISIBLE LIGHT SENSING

If internal temperature is an excellent indicator of food doneness, the surface temperature should also give a reasonable indication. Infrared sensing allows the food surface temperature to be measured (Fukuda and Miyazawa 1984) remotely by mounting an infrared sensor in the top of the oven cavity and focusing the image of the food surface on the cavity shelf on to it with a wide-angle lens. Some manufacturers have introduced models over the past two decades with only partial commercial success. Several disadvantages of the technique probably account for the inability to justify the additional cost for a sensor oven.

First, the surface temperature is known to be correlatable to doneness, i.e., center temperature, but only if food type and geometry are known. Additional information must thus be programmed into the oven. Since knowledge of food type and geometry is required, alternatively weight could be determined. With known weight and using minutes per pound, time can simply be programmed. Programming time only allows the consumer to use an inexpensive oven without the cost of an infrared sensor model.

Finally, the surface temperature sensed by the oven may not necessarily be the true temperature of the surface. The emissivity of the food surface is a measure of how effective the surface is in emitting infrared radiation. If the emissivity is equal to 1, the surface is a perfect emitter and the measured temperature is the true temperature. For most foods the emissivity is around 0.8 to 0.98. If emissivities remained constant with cooking or heating, they could be calibrated out or corrected in the microprocessor algorithm. Because emissivities may undergo large changes as the food cooks, correction is difficult, especially with foods that render fats. With fatty foods, emissivity changes unpredictably, depending on fat emanation from the surface.

With a similar concept, but sensing visible light and thus color, charring and doneness can also be detected (Tachikawa, Watanabe, and Yokoyama 1982). Visible-sensing ovens have had virtually no exposure in the marketplace. Their technical performance as yet has not been tested.

MISCELLANEOUS SENSING TECHNIQUES

Audible Sensors

Litton Microwave investigated the use of audible signatures of food cooking to determine doneness. This work resulted in a microphone-type sensor and microprocessor algorithm for the detection of popcorn popping. The popcorn oven is currently being marketed by Litton's successor, Menumaster, Inc.

Multiple Sensing

With a combination of the above technologies, it is possible to obtain a large amount of information about the cooking state of a food product. Sharp Electronics is presently demonstrating an oven using temperature, humidity, weight, and optical sensors. These sensors, in combination with a mixing bowl built into the oven, supply information to a complex algorithm that determines the state of preparation of the product. The system is known as "fuzzy logic" (Sharp 1990). The model, costing approximately $1000 in Japan, is not yet on the market in the United States.

Electric-Field Sensors

The ultimate goal of microwave oven technology is the development of the "one-button oven." With this oven, the consumer places the product to be cooked into the oven, enters if the food should be rare, medium, or well done and pushes the start button. The oven would be smart enough to sense the type of food, determine its weight, and prepare it to perfection.

The technology to develop this type of oven exists, the only question is the return on investment for the development of such an oven. All food products have characteristic microwave properties (Chapter 5), which along with geometry define how they interact with the microwave electric field within an oven cavity. This interaction distorts the electric field in the cavity and is responsible for the electric field pattern within the food itself. Working backwards, by measuring the electric-field configuration within the cavity, one should be able to determine the type of food, a rough estimate of the geometry, and thus the weight. With this information, the preparation time could be easily determined or looked up in the microprocessor memory. Electric field probe technology has recently been commercialized (Randa 1990; Wickersheim, Sun, and Kamal 1990), and the development of the ultimate oven awaits time and money.

References

Anonymous 1991. The microwave market. *Microwave World* **12**(1):10–11.

Borowski, A. and Fritz, V. 1990. Comparison of moisture determination techniques over a range of sweet corn seed moistures. *Horticultural Science* **25**(3):361.

Buck, R. 1984. *Power Controlled Microwaved Oven.* U.S. Patent 4,447,693.

Chen, D. 1976. *Food Thermometer for Microwave Oven.* U.S. Patent 3,975,720.

Fukuda, N. and Miyazawa, S. 1984. *Microwave Oven with Infrared Temperature Detector.* U.S. Patent 4,461,941.

Kobayashi, T and Kanazawa, T. 1982. *Heating Control Apparatus by Humidity Detection.* U.S. Patent 4,335,293.

Randa, J. 1990. Theoretical considerations for a thermo-optic electric field strength probe. *Journal of Microwave Power* **25**(3):133–140.

Robeson, J. 1985. Control development for an automatic cook- by-weight microwave oven. *Microwave World* **6**(1):11–13.

Sadlack, J. 1985. Temperature probes—why are they so seldom used? *The Microwave Instructor* **2**(1):2–5. Recipes Unlimited. Burnsville, MN.

Sharp 1990. Sharp News, Press Release 90–032, Osaka, Japan. Sharp Electronics Corp.

Tachikawa, H., Watanabe, M., and Yokoyama, K. 1982. *Heating Apparatus with Char Detecting and Heating Controller.* U.S. Patent 4,363,957.

Tanabe, T. and Nakagawa, H. 1982. *Microwave Oven Employing a Gas Sensor.* U.S. Patent 4,319,110.

Underwriters Laboratories 1983. *Standard for Microwave Cooking Appliance,* UL923, 2nd ed., revised May 12, 1983.

Wickersheim, K., Sun, M., and Kamal, A. 1990. A small microwave E-field probe utilizing fiberoptic thermometry. *Journal of Microwave Power* **25**(3):141–148.

4

Power for Heating and Cooking *

THE DEFINITION OF POWER

Power and energy are often confused in reference to microwave heating and cooking, especially by consumers and users of microwave ovens. Power is defined as the rate at which work is done. Work is the transfer of one form of energy into another, usually by the application of force to provide motion. Power is thus also the rate at which energy is expended or utilized.

For most people the concept of work is all too familiar. For humans and animals, their energy source is provided from the carbohydrates, protein, and fat in food eaten. Energy needed for short-term effort is stored as glycogen in the liver and muscles. Longer-term energy reserves are stored as body fat. The *calorie* is a familiar term known to dieters and is a measure of the amount of energy contained in a food. For the moment the dieter's calorie will be designated a food calorie. As activitity or work is pursued, the energy stored in the body is converted to motion, such as climbing or running. Even at rest, glycogen is metabolized or burned, producing molecular motion that is heat. By means of the human internal regulation system, this heat produces a constant, normal body temperature of 98.6°F (37°C).

A typical person may metabolize 2000 food calories of energy per day to sustain life and activity. This rate of energy consumption, equivalent to 1.4 food cal/min, is defined as power. Similarly a 750-W microwave oven utilizes energy and turns it into heat to cook food. The 750-W specification of the oven is a measure of power and is, as will be discussed, a direct measure

---

*A portion of this chapter has been adapted with permission from C. Buffler, Microwave power measurements: Impact on food processors. *Microwave World* 12(1):19–24 (1991).

of the heating rate of a product. For the microwave interaction, the energy source is the electric field in the cavity produced by the magnetron.

There is a direct equivalence between the utilization of energy in food calories and watts. One food calorie metabolized per minute is equivalent to approximately 70 W of power. As an example, a skier burning 100 food calories in a 10-min downhill run (Morehouse and Gross 1975), or 10 food calories per minute, is utilizing energy at the rate of 700 W. This number is recognized as equivalent to a typical consumer microwave oven power. Note that the first example given, metabolizing 2000 food calories per day or 1.4 food cal/min to sustain life, is equivalent to using approximately 100 W of power.

The term *food calorie* is utilized to distinguish it from the true calorie (cal) used in physics. A food calorie is actually 1000 true calories and thus is correctly designated a kilocalorie (kcal). Since engineering aspects are discussed in this book, the true calorie is used henceforth and food calories are designated kilocalories.

Absorbed Power

Chapter 1 described how the electric field of the microwave power in the microwave oven cavity interacts with food molecules to produce heat and thus a temperature rise in the material. The quantative measure of power absorbed by a load is determined by three factors: one fixed, one determined by the load, and one determined basically by the microwave oven. The absorbed power is

$$P_v = 2\pi f \epsilon_0 \epsilon'' E^2 \qquad\qquad (4\text{-}1)$$

where P_v = power absorbed per unit volume (W/m^3)

f = frequency (Hz)

E = electric field *inside* the load (V/m)

ϵ_0 = permitivity of free space (see Chapter 5 and glossary)

ϵ'' = dielectric loss factor

The microwave oven frequency f is fixed at 2.450 GHz for consumer ovens, although 0.915 GHz may be used for microwave processing. The factor ϵ'' is an electrical parameter characterizing the load that can be obtained from the literature or measured. It is discussed in Chapter 5. The electric field E *inside* the load is determined by the oven configuration, the electric or dielectric properties of the load, and the geometry of the load. The interaction of these parameters to determine the electric field distribution inside the load is extremely complex. This complexity leads to the major problem of determining the heating characteristics of microwaved products. "Rules of thumb" have been developed to give crude guidelines, and mathematical models are available to give more exact descriptions of heating patterns in foods. These concepts are discussed in detail in Chapter 6.

FACTORS AFFECTING POWER ABSORBED IN A LOAD

The heating rate and, thus, the power absorbed by a food load in a microwave oven depend on many factors. A summary of these factors follows, their order being a rough indication of their importance. Techniques for characterizing the microwave oven with regard to some of the factors are given in Chapter 8.

1. Magnetron.   The magnetron is the microwave power-generating tube. Its specification and power supply set the ultimate power available to be delivered to the food load.

The permanent magnets used in the magnetron are the cause of a major problem in consumer microwave ovens. Intrinsically, as magnets rise in temperature, the magnetic field that they produce decreases. With magnetrons, a decrease in magnetic field produces a reduction in output power. If the cooling system for the magnetron is marginally designed, the magnets heat up and the oven loses power after it is first turned on. The power drop can be as great as 20% and usually occurs over the first 5 to 10 min of use (Greenwood-Madsen and Voss 1988).

2. Power Supply.   The power supply circuit can have a ±15% influence on output power. Changes in capacitor value relate directly to oven power. Manufacturing tolerances of capacitors thus cause considerable oven-to-oven power variation in a single model line. The power transformer design is responsible for changes in power with line voltage (see Figure 2–8).

3. Load Volume.   The amount of power absorbed in a load depends on its volume with respect to the size of the oven cavity (Gerling 1987; Buffler 1990). Typical variation is illustrated in Figure 4–1. Note that power absorbed decreases rapidly as load size is decreased for a given microwave oven. Of all food-related parameters, volume has by far the largest effect.

The ratio r, defined as the power absorbed by a 500-milliliter (mL) water load to that of a 1000-mL load, is a good indicator of how an oven's power falls off with volume. For typical ovens, $r > 0.8$.

4. Load Geometry.   The microwave coupling, and thus power absorbed in identical masses of identical products, is affected by geometry. No relationship between load geometry, load orientation, and oven cavity

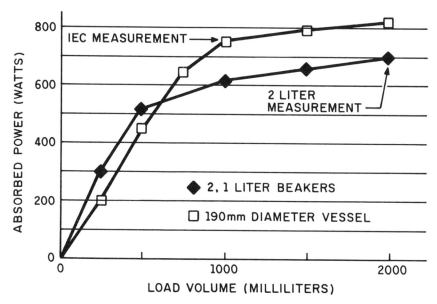

FIGURE 4-1.   Absorbed Power vs. Load; IEC and 2-L Test.

parameters has been established. It has been proposed that surface area is the key paramater to load coupling, but this hypothosis has not been verified. Further research is needed.

   5. Product or Load Temperature.   Dielectric properties, and thus absorbed power, are temperature-dependent in a complex fashion, depending on the composition or proximate analysis of the food load. The heating rate of foods whose absorption properties decrease with temperature, slow down as they heat. Ham, whose absorption increases with temperature, undergoes "thermal runaway" and may char if not carefully watched. The absorption of roast pork does not appreciably change with temperature and thus heats at the same rate as a function of temperature. The bases for these heating phenomena are described in Chapters 5 and 6.

   6. Vessel Specific Heat and Temperature.   Thermal properties of the vessel or package as well as their temperatures with respect to the load can contribute a loss of absorbed power from the load into the vessel or package. Errors introduced from this effect are usually small. For glass vessels containing 1 L of water, the error may be as high as 6%.

   7. Position of Load in Cavity.   Hot and cold spots in oven heating pattern, as described in Chapter 2, can cause ±20% variation in heating rate, particularly in small loads. For this reason, some microwave popcorn packages come with a small cardboard spacer to elevate the package if performance on the oven shelf is not satisfactory.

8. *Oven Parameters:*

a. Cavity size: Usually, the larger the cavity size compared with the load size, the less power is absorbed in the load.
b. Cavity geometry: No known relationship has been determined relating power absorbed in a load as a function of cavity geometry.
c. Cavity material: The higher the conductivity of the cavity metal, the higher the power available to the load. A stainless steel cavity usually provide 50 W more of absorbed power to a 1-L load than will a similar, painted, cold-rolled steel cavity.
d. Microwave feed system: No known relationship has been established. Stirrer and feed design is critical for obtaining high coupling efficiency to the load and for uniform heating characteristics.

Other oven-related considerations affecting power are described in Chapter 2.

POWER MEASUREMENT TECHNIQUES

Once power has been absorbed in a load, as indicated by Equation 4-1, the ensuing molecular motion produces a temperature rise. The rate of temperature rise is determined by the absorbed power as well as physical and thermal properties of the load. The temperature rise ΔT of any medium and the total power P absorbed in it, are related (in SI units) as follows:

$$P\ (W) = P_v V$$
$$= 14.7 \frac{V\ (\text{m}^3) \times \rho\ (\text{kg/m}^3) \times C_p\ (\text{J/kg K}) \times \Delta T\ (\text{K})}{t\ (\text{s})} \tag{4-2}$$

or, in more conventionally used units,

$$P\ (W) = 69.8 \frac{V\ (\text{L}) \times \rho\ (\text{g/mL}) \times C_p\ (\text{cal/g}°\text{C}) \times \Delta T\ (°\text{C})}{t\ (\text{min})} \tag{4-3}$$

where V = total sample volume
ρ = load density
C_p = load heat capacity (specific heat)
t = time that microwave power is applied.

Note that equations with English units are not used in the food industry, even in the United States.

Equation 4–2 simplifies as follows for water (70 is used as an approximation for the multiplication factor):

$$P \text{ (W)} = 70 \frac{V \text{ (L)} \times \Delta T \text{ (°C)}}{t \text{ (min)}}$$

(4–4)

For example, consider 1 L of water in a microwave transparent dish placed in a microwave oven for 1 min. If a temperature rise of 9°C was measured, then from Equation 4–4 630 W were absorbed. Note that 630 W refers to power, that is, the rate at which energy is being absorbed. Power thus is *not* the total amount of absorbed energy, but the rate that energy is being supplied to the load. It is therefore incorrect to refer to a microwave oven as having 630 W of energy.

Equation 4–4 has been used for decades for measuring microwave oven power. This measurement has a checkered history. Originally 1 L of water was the volume established as an informal standard. In the late 1970s, Japanese manufacturers realized the significance of the load volume effect illustrated in Figure 4–1. By using 2 L for their power measurement, they were able to increase the advertised power on virtually all their ovens. Their introduction of the 2-L test confused the marketplace, but in time U.S. manufacturers all switched to the 2-L test procedure (IMPI 1989). Parallel with U.S. efforts, the International Electrotechnical Commission (IEC) was establishing yet another measurement technique (IEC 1988), using 1 L of water in a 190-mm-diameter vessel (Corning) but cooled to approximately 10°C (50°F). In many instances the IEC method produces a higher power result than the 2-L test, even though 1-L of water is used in the IEC test (Figure 4–1). The test water is cooled to 10°C (50°F), to eliminate lower readings caused by power lost in heating the vessel. In addition, the cold water has a higher microwave absorption than room temperature water, resulting in a higher measured power.

In 1991, one oven manufacturer introduced the IEC tests into part of their product line (Sharp 1990). Many other manufacturers soon followed, resulting in further widespread confusion (Buffler 1991). Consumers purchasing IEC-labled ovens using recipe books and products with instructions, based on the 2-L tests, could experience poor performance. The history of microwave power measurement points to the necessity for strong self- or government regulation.

An alternative approach to power-labeling microwave ovens is presently under consideration. Instead of rating by power or wattage, an oven could simply be designated "high" or "low." The differentiation could be accomplished by measuring the time to boil a cup of water (Buffler 1988). Microwave instructions could be developed with these categories; and consumers could determine their own oven's characteristic, independent of the rating used by the manufacturer. A cup test designed for the consumer has been developed by IMPI, which distinguishes high- and low- power ovens referenced to a dividing line of 600 W as measured by a 2-L test. Measurement techniques for the 2 L, IEC, and a one-cup (U.S.) time-to-boil test (IMPI 1991) are given in Appendix 5.

MICROWAVE OVEN EFFICIENCY

The model, serial number, and voltage and current specifications for a microwave oven are listed on a plate at the rear of the oven. For an oven with a current (I) specification of 13.3 amperes (A) at a rated line voltage (V) of 120 V, the power drawn from the wall receptacle is $P = IV$, or 1596 W. This power is known as the oven input power. If a bowl test determined that the oven output power was 700 W, the efficiency of power transfer from the line receptacle to the water would be 700/1596 = 0.44 or 44%. This value might seem surprisingly low, since microwave ovens are thought to be very efficient compared with conventional ranges. Actually, efficiencies around 50% are typical for all microwave processes, whether for an oven or an industrial system.

Figure 4–2 illustrates the power flow in a typical microwave oven. Most of the power is dissipated as heat from the magnetron, so a cooling fan or blower is required in all microwave ovens. A smaller amount of heat is generated by the transformer, light bulb, and electronic circuitry if applicable. Once the microwave power is generated, the energy transfer to the load can be quite efficient, as high as 85% to 95% for large-volume loads. The high efficiency of microwave cooking comes from a comparison of cooking foods such as corn on the cob or baked potatoes. Conventionally, a large pot of water must first be boiled to cook corn, requiring a large amount of energy. In a conventional oven used to bake potatoes, 35 lb of steel must first be heated as well as the few potatoes. The microwave oven, on the other hand, only requires 3–4 min cooking per potato or cob of corn, a much more efficient process than conventional cooking.

The Department of Energy (DOE) is considering mandating higher efficiencies for microwave ovens sold. Efficiencies could be increased by approximately 150 W by increasing the quality of transformer and cavity materials. However, microwave oven magnetrons are typically operated only 100 h/yr, providing a realizable savings of 15 kilowatt hours (kWh).

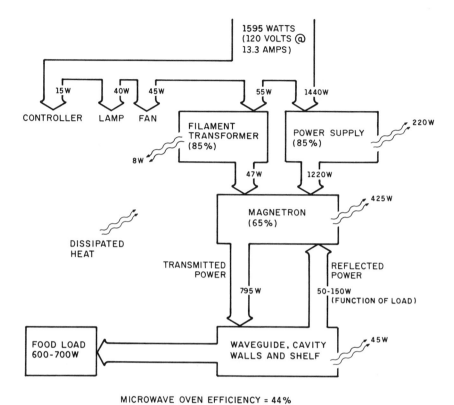

MICROWAVE OVEN EFFICIENCY = 44%

FIGURE 4-2.   Efficiency of Microwave Oven.

At 13 cents/kWh a saving of $1.95 per year does not justify the increase in selling price.

References

Buffler, C. 1988. A simple home test to determine microwave oven power output. *Microwave World* 9(2):5-8.

Buffler, C. 1990. An analysis of power data for the establishment of a microwave oven standard. *Microwave World* 11(3):10-15.

Buffler, C. 1991. Microwave power measurements: Their impact on food processors. *Microwave World* 12(1):19-24.

Corning. Crystallizing Dish Model 3140-190. Corning Glass Works, Corning, NY.

Gerling, J. 1984. *Correlation Data and Analysis.* Report #84-001. Gerling Laboratories, Modesto, CA.

Gerling, J. 1987. Microwave oven power: a technical review. *Journal of Microwave Power* 22(4):199-207.

Greenwood-Madsen, T. and Voss, W. 1988. Microwave oven power measurements. *Electromagnetic Energy Reviews* **1**:27–31.

IEC 1988. *Methods for Measuring the Performance of Microwave Ovens for Household and Similar Purposes.* CEI IEC 705, 2nd ed. Bureau Central de la Commission Electrotechnique Internationale, Geneva, Switzerland.

IMPI 1989. A guideline for power output measurement of consumer microwave ovens. *Microwave World* **10**(5):15.

IMPI 1991. *Time to Boil Test.* News Release. International Microwave Power Institute, Clifton, VA.

L. Morehouse and L. Gross 1975. *Total Fitness in 30 Minutes a Week.* New York: Pocket Books.

Sharp 1990. *Carousel II Microwave Ovens,* Catalog MW-06–020, Sharp Electronics Corp., Mahwah, NJ.

5

Dielectric Properties of Foods and Microwave Materials

DEFINITION OF DIELECTRIC PROPERTIES

Because microwave radiation is part of the electromagnetic spectrum, it behaves very much like the more familiar visible light radiation we experience daily. Similar to the laws of optics, materials interact with microwaves in three ways. First, they reflect microwave radiation impinging on them. Second, they transmit microwaves that have entered into them. Finally, they absorb some of the microwave energy being transmitted through them.

Mathematical equations developed by Maxwell (1873) predict the complete behavior of electromagnetic radiation's interaction with matter for any type of material in any geometry. In order to do this, two pairs of parameters describing the material are required. These pairs are known as the complex* permeability and the complex permittivity.

The permeability and permittivity parameters are defined as

$$\mu = \mu' - j\mu'' \tag{5-1a}$$

$$\epsilon = \epsilon' - j\epsilon'' \tag{5-1b}$$

where μ = complex permeability
μ' = magnetic permeability
μ'' = magnetic loss factor

---

*The term *complex* refers to a type of algebra that uses imaginary numbers associated with $j = \sqrt{-1}$. The first term, a, in a complex number of the $a + jb$, is the real portion, and the second term, preceded by j (or i in mathematics), is the imaginary portion. Discussion of complex numbers can be found in any college algebra or complex variable text. Complex algebra simplifies the calculation of the sinusoidal behvaior of electromagnetic waves. The technique is described in advanced electromagnetic texts.

ϵ = complex permittivity

ϵ' = permittivity

ϵ'' = dielectric loss factor

Equations 5-1 can represent the absolute values of permeability or permittivity of a material or, as is commonly done, the values referenced to free space. In the latter case the terms are called *relative* permeabilities or permittivities. Common usage, followed in this book, uses relative values. Under certain circumstances the absolute values of permeability and permittivity of free space must be designated and will be so noted. Further details of the definitions are given in Appendix 2.

No foods magnetically interact with microwave radiation. Thus, for foods Equation 5-1a reduces to $\mu = 1$, the permeability referenced to that of free space (see glossary for value). For magnetic materials such as ferrites, often used in susceptors and browning dishes, the magnetic interaction is appreciable. For these materials the magnetic interaction accounts for the observed heating. Also, in instances where microwaves interact with a metallic boundary, such as a microwave oven cavity, permeability needs to be considered. If the metal is at all magnetic, such as with some types of stainless steel, excessive losses can take place. In this book, only electrical interaction is considered, and only measurement techniques for dielectric permittivity are addressed.

To further confuse the issue, there is as yet no official nomenclature or symbolism to describe dielectric parameters. The definition of Equation 5-1b is as close to a consensus as is available. The term *permittivity* is not in common usage in the United States in the field of food science. Instead, the term *dielectric constant* is used. Thus, in Equation 5-1b, ϵ is called the complex dielectric constant, ϵ' the dielectric constant, and ϵ'' the loss factor. Again, these values are all relative to the values of free space. This terminology is used throughout the book. If any confusion between magnetic and dielectric loss is possible, the term *dielectric loss factor* should be used.

The values of ϵ' and ϵ'' of a food material play a critical role in determining the interaction of the microwave electric field with the material. A "map" of foods plotted against their dielectric parameters was introduced by Bengtsson and Risman (1971). Appendix 3 presents dielectric values, and Figure 5-1 shows a "map" for common foods.

REFLECTION, ABSORPTION, REFRACTION, AND TRANSMISSION

Reflection

To understand how microwaves interact with materials, we examine a simple case, a plane wave impinging on an infinite slab of material. A plane wave is defined as microwave radiation whose electric field directions in

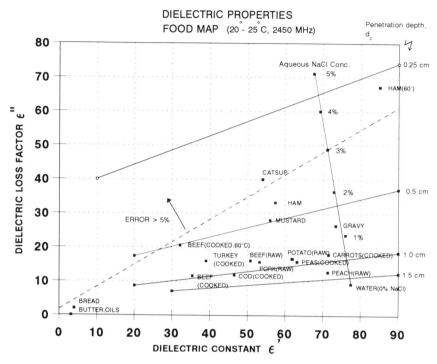

FIGURE 5-1.   Map of Dielectric Properties of Foods (Buffler and Stanford 1991).

space are all parallel. Such is the case for an electromagnetic wave far from its generating antenna, as illustrated in Figure 1–5c, or as exists in an empty oven cavity with the microwaves bouncing from wall to wall.*

For small loads in a microwave oven, the assumption of plane waves impinging on the food may have somewhat more validity than for large loads, particularly if the load is thin and flat. For larger loads the assumption is inaccurate but still gives a basis for visualization.

The term *infinite slab* indicates a sample of material that extends with no bounds in two directions but has a finite thickness. This assumption eliminates the need to consider what happens at edges or corners.

---

*In actuality, in a region bounded by metallic walls such as a cavity or waveguide, true plane waves cannot exist since the electric field parallel to the metallic wall boundary must be zero. Two counterreflecting plane waves must exist, bouncing off opposite walls, to fulfill the boundary conditions. These waves make up a "guided wave" with a wavelength longer than the free-space wavelength. For illustrative purposes the plane-wave concept is used in this book. Full treatment of guided waves is given in many electromagnetic theory texts, such as Marcuvitz (1986).

For microwave radiation incident upon a slab from a direction perpendicular to its surface, a fraction of the energy is reflected from the surface, P_r, depending on its complex dielectric constant ϵ. The principal determining factor for the magnitude of reflection, is from the dielectric constant ϵ' of the material. Errors due to neglecting ϵ'' are less than 5% for virtually all foods, as indicated by the 5% line in Figure 5-1. Neglecting the loss factor, an approximate relation for the fraction of microwave power reflected from an infinite-slab food surface is

$$P_r = \left(\frac{\sqrt{\epsilon'} - 1}{\sqrt{\epsilon'} + 1}\right)^2$$

(5-2)

Note that $\sqrt{\epsilon'}$ is equivalent to the well-known optical index of refraction η. $\sqrt{\epsilon'}$ can thus be thought of as the microwave index of refraction in future discussion.

Absorption

If the fraction of power reflected is P_r, the amount transmitted into the medium, P_o, is given by $P_o = 1 - P_r$. Once the microwave energy P_o enters the food, it propagates internally, perpendicular to the surface, toward the opposite face of the slab. If the material is microwave absorptive or lossy, the propagating energy decreases as it traverses the slab as more and more of the energy is absorbed (Figure 5-2). The parameter that measures the microwave absorptivity of a material is the loss factor ϵ''. The loss factor is zero for a nonabsorbing medium, and increases to 20 to 30 for highly absorbing foods such as ham and salted products.

The fundamental equation for microwave power absorption is Equation 4-1. Converted to units more commonly used, the power absorbed becomes

$$P_v = 5.56 \times 10^{-4} \times f \epsilon'' E^2$$

(5-3)

where P_v is the power absorbed per unit volume (W/cm³)

$f =$ frequency (GHz)

$\epsilon'' =$ relative dielectric loss factor

$E =$ electric field (V/cm).

The multiplying constant takes into consideration the dielectric constant of free space and adjusts for the non-SI units used.

Thus, as microwave energy propagates through the slab, the power at any point, P, and the power dissipated per unit volume, P_v, decrease. For

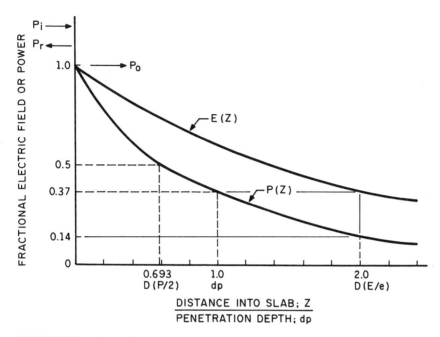

FIGURE 5-2.   Penetration Depth of Electric Field and Power into a Material.

materials with high loss factors power decreases rapidly and the microwave energy does not penetrate deeply. For lower-loss materials the microwave energy may penetrate extensively. A parameter, designated penetration depth, will be defined later and is extremely important in determining how microwaves interact with materials.

Refraction

For microwave radiation incident on a surface at an oblique angle α, the reflected wave is at the same angle (Figure 5-3). The transmitted wave is refracted toward a line perpendicular to the surface (normal direction) at an angle β. The magnitude of β is given by Snell's law:

$$\sin \beta = \frac{\sin \alpha}{\sqrt{\epsilon'}}$$

(5-4)

Equation 5-4 is approximate; a correct equation includes the loss factor. The error introduced by using only ϵ' in Equation 5-4 is less than 5% for

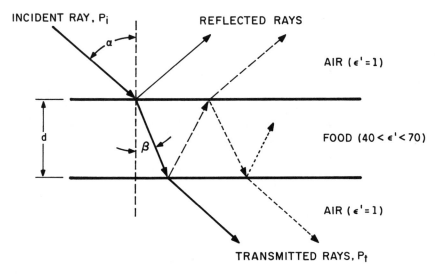

FIGURE 5-3.   Reflection, Absorption, and Transmission of Microwaves in a Dielectric Medium.

virtually all foods. The 5% error boundary line on Figure 5-1 also holds approximately for refraction.

Examining the initial refracted ray shows that if the absorption of the medium is not too great, a portion of the microwave power reaches the opposite side of the slab. At this interface a portion of the wave is reflected internally according to Equation 5-2, and a portion is transmitted out the opposite side. This reflection process continues until there is essentially no remaining microwave power at an interface; that is, the power has been essentially all absorbed during the multiple reflections.

Transmission

The power reflected from the surface by an oblique incident wave shown in Figure 5-3 consists of the sum of the initially reflected wave plus all the waves emanating from that surface produced by the internal multiple reflections.

Additionally, there is some power that escapes from the opposite side of the slab during these multiple reflections. The total power transmitted through the product, P_t, is the sum of the power emanating at each internal reflection. The same analysis holds for perpendicular incidence, $\alpha = 0°$, previously discussed. Energy, and thus power, is conserved; in other words, it must all be reflected, absorbed, or transmitted through the slab. The fraction of power transmitted is thus: $P_t = 1 - P_r - P_a$.

Wavelength Inside a Medium

From Chapter 1 and Equations 1–2 it is seen that the wavelength of propagating microwave radiation is a function of the velocity of propagation and of the frequency. As with visible light, the velocity of propagation is slowed in the medium by the index of refraction $\eta = \sqrt{\epsilon'}$ so that Equation 1–2a becomes

$$\lambda_m = \frac{c/\sqrt{\epsilon'}}{f}$$

(5–5)

Thus, for a fixed microwave frequency f the wavelength inside the medium, λ_m, is reduced from the free-space wavelength λ_0. At the microwave oven frequency of 2.45 GHz the wavelength in the medium is $12.8/\sqrt{\epsilon'}$ cm or $11.8/\sqrt{\epsilon'}$ in.

Because microwave radiation experiences multiple internal reflections, the amplitudes of all the internal rays add or subtract at all positions within the slab. The phenomenon is the same as that causing hot and cold spots within an oven cavity, as described in Chapter 2. The distances between maxima and minima of the internal reflected waves depend on λ_m. The amplitudes of the waves depend on the magnitude of the reflections at each interface as well as the amount of power absorbed at each point in the sample. Fortunately this simple model can be solved directly by using Maxwell's equations to show the variations of power density within a slab (Ayappa et al. 1991). Without reverting to the mathematical solution, it is possible to say that the reflected signal as well as the absorbed power and transmitted signal exhibit sinusoidal like variation as the slab thickness is changed. This variation has a period given by the wavelength, λ_m, within the material given by Equation 5–5.

THE DIELECTRIC SPECTRUM

Chapter 1 described the two mechanisms by which the microwave electric field was converted to heat within a material. The first, ionic, comes from a linear acceleration of ions, usually from salts, within a food. The second is the molecular rotation of polar molecules, primarily water, as well as weaker interactions with carbohydrates and fats.

To understand how different materials have different dielectric properties and to understand the temperature and frequency behavior of the microwave interaction, we need some knowledge of the fundamental physics of these two absorption mechanisms.

Ionic Interaction

As the dissolved charged particles in a food or material, usually ions, oscillate back and forth under the influence of the microwave electric field, they collide with their neighboring atoms or molecules. These collisions impart agitation or motion, which is defined as heat. Materials with mobile ions are *conductive,* in that the movement or flow of charged particles is defined as the conduction of electricity. The more available conducting ions a material has the higher is its electrical conductivity. Since the microwave absorption of a material depends on the number of ions it can interact with, the microwave absorption of such a material increases with its conductivity.

Interestingly, the electromagnetic absorption of ionic materials can be characterized by their conductivity under the influence of a constant (dc or direct current) electric field.

By examining the wave equation derived from Maxwell's equations, one sees that the conductivity contribution to the power absorbed by a material acts as if it were part of the losses described by the dielectric loss factor. The details derived in most electromagnetic texts, and reviewed by Geyer (1990), show that the ionic or conductivity contribution to the loss factor, ϵ'', is

$$\epsilon_\sigma'' = \frac{\sigma}{2\pi f \epsilon_o} \tag{5-6a}$$

In SI units σ represents the dc conductivity* in siemens (commonly called mhos), f is the frequency in hertz, and ϵ_o is the permittivity of free space $(8.85 \times 10^{-12}$ F/m). In commonly used units, with σ in mmhos/cm and f in GHz, Equation 5-6a becomes

$$\epsilon_\sigma'' = \frac{1.80\sigma}{f} \tag{5-6b}$$

---

*Most conductivity measurements are made at frequencies in the low-megahertz range in order to reduce errors caused by the measurement electrodes when measuring at dc. Errors introduced at microwave frequencies using these values are negligible.

Thus, at 2.45 GHz, a salted food with a conductivity of 11 mmhos produces a contribution to the loss factor of $\epsilon_\sigma'' = 8.08$.

The loss factor in such a conductive material thus increases monotonically with decreasing frequency, according to Equations 5–6, and becomes infinite at zero frequency (dc) (Figure 5–4). The apparent anomalous blowup of the loss factor at low frequencies might at first seem strange. The reason is that under the influence of a dc electric field, ions move continuously. This situation is unlike polar molecules, which stop their rotation as soon as they are aligned with the field. The key point is that the power absorbed by the moving ions does not increase without limit as the frequency is lowered. If one substitutes Equation 5–6a into Equation, 4–1 to calculate the absorbed power, one sees that the frequency terms cancel and $P_v = \sigma E^2$, independent of frequency. This fact is not sufficiently stressed in the literature, where the impression is given that the power absorbed increases without bound as the frequency is decreased.

The temperature dependence of this conductivity contribution depends on the temperature dependence of the dc conductivity. Typically the ionic conduction increases with temperature, giving the behavior of ϵ_σ'' shown in Figure 5–4.

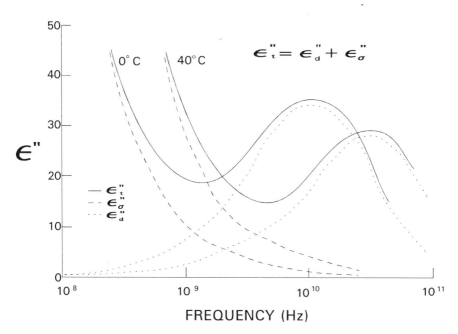

FIGURE 5–4.   Dielectric Loss Factor vs. Frequency   (Adapted from Roebuck, Goldblith, and Westphal 1972).

Polar Interaction

If a material of polar molecules, such as water, is exposed to a fixed or static electric field, the molecules rotate to orient themselves in the direction of the field. The strength of the separated charge at the ends of the molecules is defined as the dipole moment, and is a measure of the strength of interaction with the field. The dipole moment is also a measure of the dielectric constant ϵ'. A symmetrical molecule, with no dipole moment, is said to be nonpolar and does not react with an electric field. If the frequency of the electric field is increased, the molecules rotate, following the alternating reversals of the electric field. As long as the polar molecule can rotate freely, the dipole moment is fully effective and the dielectric constant remains constant with frequency. Because polar molecules interact with other molecules in the material, they transfer their motion, imparted from the electric field, to the entire sample as heat. This intermolecular interaction also acts as a retarding or damping effect on the molecule itself. It is thus possible to visualize the polar molecules as if they existed in a highly viscous fluid similar to the damping fluid in a shock absorber.

If the frequency of the electric field is increased still further, the molecules continue their attempt to rotate with the field, but are impeded more and more by the viscous damping. The molecules can no longer rotate fully, the dipole moment is less effective, and the measured dielectric constant decreases.

The dielectric loss or absorption behaves differently. At very low frequencies the dipole follows the field freely, and little energy is transferred to the surrounding molecules, so little absorption occurs. As the frequency increases, molecular impediment increases and more and more energy is lost to the surrounding molecules, which can be thought of as the effective viscous medium. A maximum absorption is reached, and as the frequency is raised still further the dipole can no longer move in response to the rapidly oscillating field. The dipole, in this situation can no longer transfer energy to its surroundings, and the absorption decreases toward zero.

The maximum absorption point is defined as the *relaxation frequency* or *critical frequency* f_c. A reciprocal function of this frequency is called the *relaxation time* $\tau = 1/2\pi f_c$. The relaxation time is the time it takes an agitated molecule to relax back to 36.8% of its original condition once the stimulus has been removed.

A plot of the loss factor, as a function of frequency for the polar contribution ϵ_d'', is shown for water in Figure 5-4. Note that the relaxation frequency is approximately 18 GHz. The microwave absorption for 915 MHz and 2450 MHz thus takes place below this critical frequency.

The relaxation or critical frequency of a material is related to its structure. Many organic liquids, including cooking oils (Pace, Westphal, and Goldblith 1968), are polar and have relaxation frequencies in the low-megahertz range. These molecules can be thought of as inertia-bound and cumbersome to move. Thus, very little change in loss factor is seen above this low frequency relaxation frequency up to the measurement limit of most dielectric measurement equipment (i.e., 20 GHz). Liquid water is much freer to move and thus has a higher relaxation frequency. Solids not containing ions, such as ice and plastics, cannot strictly be thought of as polar. These molecules are locked into place by their structure and are unable to move easily. They thus are unable to participate readily in dielectric absorption and, consequently, have low values of dielectric constant and loss factor (see Appendix 3). The much lower values of dielectric constant and loss factor of ice compared with water constitute a severe problem for microwave thawing, as discussed in Chapter 6.

The temperature and frequency dependence of the dielectric properties of polar molecules such as water was first modeled by Debye (Debye 1929). Early work on dielectric properties has been described by von Hippel (1954a,b). Excellent recent reviews are Ohlsson and Bengtsson (1975), Mudgett (1985; 1990), and Geyer (1990).

Combined Mechanisms

The simplest model for the dielectric properties of foods is the distributive model. Here, the dielectric properties of each constituent of the food are added according to their volume fractional makeup of the total product. The model assumes that the various food constituents are distributed reasonably uniformly throughout the product. Figure 5-4 shows the total dielectric loss factor for a 0.5 molar (M) aqueous solution of water at two temperatures. Note that the total loss factor, ϵ_t'', is the simple sum of the ionic and polar contributions: ϵ_σ'' and ϵ_d''.

The synergistic effect noted among some mixtures of solutions (Roebuck, Goldblith, and Westphal 1972; Engelder and Buffler 1991) cannot be easily explained by the distributive model. For example, mixtures of glycerol and water and ethanol and water show a maximum in loss factor at 3 GHz at concentrations of 50% and 22%, respectively, at values that are considerably higher than the loss factor of either constituent. These observations are understood when it is realized that water has a critical frequency around 18 GHz, and the heavier organic liquids have a critical frequency quite low in the megahertz range. If the dielectric properties were simply additive, the loss factor vs frequency would be a bimodal distribution with the amplitudes of the peaks varying as a function of concentration. Instead, an interaction takes place between the two molecules, forming a solution with a loss factor peak that depends on both molecules. As the polar liquid is added to water, the critical frequency of the mixture decreases until it reaches the liquid frequency in the megahertz range. Then ϵ'', measured at any frequency, at first increases as a function of concentration, reaches a maximum when the critical frequency equals the measurement frequency, and then again decreases: the lower the measurement frequency, the higher the concentration required to reach the maximum ϵ''. This analysis is well borne out by the measurements on glycerol and ethanol mixtures (Roebuck, Goldblith, and Westphal 1972) and methanol mixtures (Engelder and Buffler 1991). Because of the interactive nature of many food combinations, it is extremely important to obtain information about the full frequency spectrum of dielectric properties of food system constituents. Judgments can then be made about their performance in differing concentrations, and insight can be gained on their temperature behavior.

Finally, an area needing much research is the dielectric properties of two-phase systems such as emulsions, whips, and foams. It is well known that the dielectric behavior of particles of one dielectric property imbedded in a substrate of another behave very differently than a distributive mixture of both. Fricke (1955) developed a model for randomly oriented oblate spheroids suspended in a continuous medium. This model was extended by Dendgett, Wang and Goldblith (1974) and is expected to be used successfully to model two-phase food systems, but to date there is very little literature reporting such studies.

Temperature Dependence of Dielectric Properties of Foods

Research has not yet produced an analytical solution to the temperature dependence of the parameters that enter into the Debye equations. Only qualitative explanations as well as empirical data exist. Bengtsson and Risman (1971) have published data on dielectric properties of foods as a function of temperature. These data are reproduced as Figures 5-5 and 5-6. The data are for 3 GHz but are representative of values at 2.45 GHz. Note the decrease in ϵ'' with temperature of pure water, as expected from Figure 5-4. As the food becomes saltier, the ionic effect comes into play, reducing the drop-off of ϵ'' with temperature. For ham, the ionic contribution dominates and the loss factor increases rapidly with temperature.

Continued development of models incorporating temperature dependence of dielectric properties would facilitate the predictions of food systems for use by food scientists or product developers.

Other important works, containing copious references, on dielectric theory, measurement techniques, and data tabulation have been published. Pi-

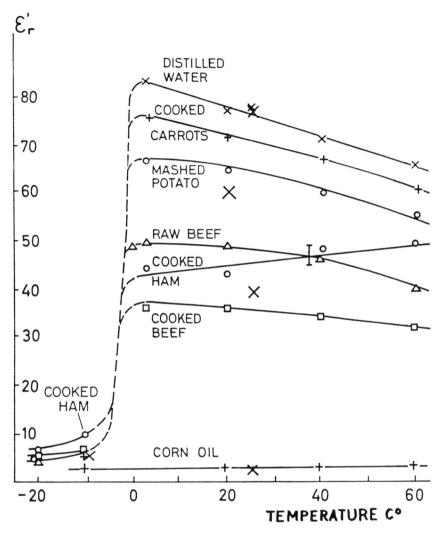

FIGURE 5-5.   Dielectric Constant of Foods vs. Temperature   (Adapted from Bengtsson and Risman 1971).

oneering work was done by von Hippel (1954*a,b,c*) at the MIT Laboratory for Insulation Research. These books are out of print but should be sought by serious workers in the field. Buckley and Maryott (1958) at the National Bureau of Standards have tabulated data on liquids. Nelson (1991), from the U.S. Department of Agriculture, and Tinga (Tinga and Nelson 1973) have tabulated dielectric information on agricultural as well as other mate-

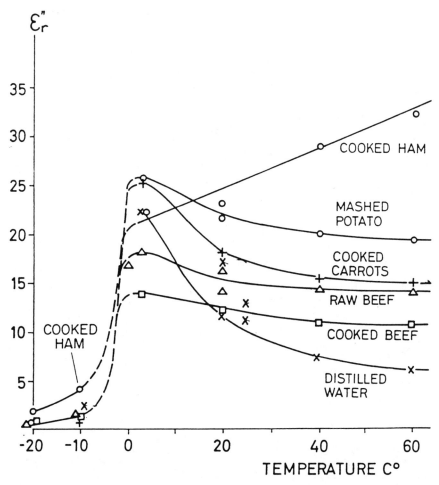

FIGURE 5-6.   Loss Factor of Foods vs. Temperature   (Adapted from Bengtsson and Risman 1971).

rials. Ohlsson and Bengtsson (1975) and Kent (1987) have published data on foods, El-Rays and Ulaby (1987) on vegetative materials.

PENETRATION DEPTH

As discussed, the more microwave absorptive a material is (i.e., the higher the loss factor ϵ''), the less deep microwave energy will penetrate into that material. It is very instructive to define a parameter that indicates how far into a material a microwave penetrates before it is reduced to a certain fraction of its initial value. This parameter, defined as penetration depth, is a function of both dielectric properties of a material, ϵ' and ϵ''. Knowledge of penetration depth serves as a guideline to the heating effectivity of a food and is therefore an important parameter that must be known by food product developers. A full discussion of microwave heating is in Chapter 6.

Unfortunately, the term *penetration depth* has three common, slightly different definitions. Over the years charts and tables of published values of foods and materials have sometimes been lax in including the exact definition used. To make matters worse, many publications, as well as tabulated values, have used incorrect formulas, thus promulgating confusion among readers and users.

Definitions

Part of the present confusion in the literature arises from the definition of the term *wave* and the specification of the fraction of that wave that remains at that given penetration depth. Three definitions are currently in common use in the literature:

Power Penetration Depth, D(P/e) or d_p

The most useful concept for the microwave food professional is the concept of power penetration, since the heating of a product depends directly on the power available at a given position in the product. The equation describing the power at any point within an infinite slab of material for an incident plane wave is

$$\frac{P(z)}{P_o} = e^{-z/d_p} \tag{5-7}$$

where $P(z)/P_o$ = fraction of power remaining as a function of distance into the material

e = Napierian logarithm base, a mathematical constant = 2.718.

The units of distance z and penetration depth d_p are arbitrary, since they occur as a ratio. Centimeters are most commonly used in the literature. Equation 5-7 is plotted in Figure 5-2.

The penetration depth, d_p, is thus defined as the depth at which the microwave power has decreased to $1/e$ or 36.8% of its original power. The "original value of power" is defined as that power which has entered the sample, P_o—the incident power P_i, minus the reflected power, P_r. The $1/e$ power definition of penetration depth, has gained international acceptance in the scientific community.

Electric Field Penetration Depth D(E/e)

The fundamental equations describing electromagnetic waves, and thus microwaves, are written in terms of the electric field E. The electric field penetration depth is defined as the depth at which the electric field has diminished to $1/e$ of its original value. Since the electric field is the square root of the power, the exponent in Equation 5–7 is multiplied by $1/2$. The electric field penetration depth, i.e., the value of z for which the exponent equals 1 is then $2d_p$. This definition is commonly used by microwave engineers, but not predominantly used by food scientists.

Half-Power Penetration Depth D(P/2)

Another commonly used definition of penetration depth is the half-power penetration depth. This definition is the most easily understood conceptually, being simply the distance into a material that microwaves must penetrate before the microwave power is reduced to half its original value. The mathematical simplicity of d_p, given by Equation 5–7, has won out over the conceptual simplicity of half-power penetration depth. The relationship between the three definitions is presented in Table 5–1. Figure 5–2 shows the power and electric field decrease inside a semi-infinite slab of material for an impinging perpendicular incident wave.

Penetration Depth as a Function of Dielectric Parameters

The penetration depth of a material depends strongly on ϵ' and ϵ''. Many references give an incorrect formula for the $1/e$ power penetration depth, the correct formula is

TABLE 5-1   Relationships Among Different Definitions of Penetration Depth

		To Obtain		
Multiply		D(E/e)	D(P/e)	D(P/2)
$D(E/e) = 1/a$	by	1	0.5	0.347
$D(P/e) = d_p$	by	2.0	1	0.693
$D(P/2)$	by	2.885	1.443	1

$$d_p = \frac{\lambda_o\sqrt{2}}{2\pi} \{\epsilon' \ [\sqrt{1 + (\epsilon''/\epsilon')^2} - 1]\}^{-\frac{1}{2}}$$

$$(5\text{-}8)$$

For 2450 MHz, $\lambda_o \sqrt{2}/2\pi = 1.085$ in (2.76 cm); for 915 MHz the value is 2.91 in (7.38 cm). Slightly different forms of Equation 5-8 exist in the literature. Always be sure the equation is correct before making calculations! For low values of ϵ'', Equation 5-8 can be dramatically simplified to

$$d_p = D(P/e) = \frac{\lambda_o\sqrt{\epsilon'}}{2\pi\epsilon''}$$

$$(5\text{-}9)$$

Even though this approximation should hold only for low-loss materials ($\epsilon'' << 1$), such as fats and oils, it is remarkably accurate for most food materials. The accuracy is better than 5% for most commonly encountered foods, as shown on the food map (Figure 5-1). A comparison of power penetration depths for various materials is given in Appendix 3. An excellent detailed review of penetration depth is given by Metaxas (1985).

MEASUREMENT OF DIELECTRIC PROPERTIES

Waveguide and Coaxial Transmission Line Method

Early efforts characterizing dielectric properties of materials was carried out at the Massachusetts Institute of Technology (Roberts and von Hippel 1946; von Hippel 1954b). The values of ϵ' and ϵ'' were derived from transmission line theory, which indicated that these parameters could be determined by measuring the phase and amplitude of a reflected microwave signal from a sample of material placed against the end of a short-circuited transmission line, such as a waveguide or a coaxial line. For a waveguide structure, rectangular samples that fit into the dimensions of the waveguide at the frequency being measured are required. For coaxial lines, an annular sample needs to be fabricated. The thickness of the sample should be approximately one-quarter wavelength within the sample. Preparing an optimal sample, therefore, requires guessing the dielectric constant of the material being measured so that the wavelength can be determined. Typical thicknesses at 2450 MHz range from 0.2 in (0.5 cm) for foods to 0.75 in (1.9 cm) for fats and oils.

The disadvantage of fabricating annular samples is outweighed by the fact that a single coaxial line can be used over a very wide range of microwave frequencies. Waveguides, on the other hand, have narrow operating ranges, requiring many differing sizes as the frequency of measurement is changed. Also, quarter-wavelength estimations must be corrected for the effect of the guide wavelength. (See Marcuvitz 1986, p. 62.) For single-frequency measurements, the waveguide technique is usually chosen.

The dielectric parameters can be easily and inexpensively obtained by the transmission line technique, particularly if one utilizes a slotted line and standing-wave indicator (Nelson, Stetson, and Schlaphoff 1974). This equipment is presently available on the used microwave equipment market (cf. Appendix 4 for suppliers). A more sophisticated implementation of the technique utilizes a swept-frequency network analyzer, where the impedance is measured automatically as a function of frequency (Hewlett Packard 1985).*

The transmission line technique is somewhat cumbersome because the sample must be made into a slab or annular geometry. At 2450 MHz the sample size is somewhat large, particularly for fats and oils. Commonly available waveguide test equipment for 2450 MHz is designated WR-284. For measurements at 915 MHz, only the coaxial line technique is practical due to the large size of waveguide required. Liquids and viscous-fluid-type foods can be measured with this method by using a sample holder at the end of a vertical transmission line.

An additional problem encountered with WR-284 is that many components are not calibrated below 2600 MHz. Extreme care should be taken when setting up a WR-284 system at 2450 MHz. Recalibration of components may be necessary. Alternatively, less readily available waveguide equipment, such as WR-340 or WR-430, can be used. Characteristics of commonly used waveguide are given in Appendix 6.

Open-Ended Probe Technique

A method that circumvents many disadvantages of the transmission line measurement technique was pioneered by Stuchley and Stuchley (1980). The technique calculates the dielectric parameters from the phase and amplitude of the reflected signal at the end of an open-ended coaxial line inserted into a sample to be measured. Care must be exercised with this technique because errors are introduced at very low frequencies and at very high frequencies, as well as for low values of dielectric constant and loss factor. The technique is valid for 915 and 2450 MHz, for materials with

---

*This reference describes both the reflection technique for making dielectric measurements and a combined reflection and transmission technique that allows permittivity and permeability to be calculated.

loss factors greater than 1. Interpretation for lower-loss materials such as fats and oils must be treated with caution. Typical open-ended probes utilize 3.5-mm- (0.138-in-) diameter coaxial line. For measurement of solid samples, probes with flat flanges may be utilized (Hewlett Packard 1991).

The open-ended probe technique has been successfully commercialized and software and hardware are available (see Appendix 4). Figure 5–7 shows an open-ended probe system. The practical aspects of this technique have been described in detail by Engelder and Buffler (1991).

Cavity Perturbation Technique

The measurement errors intrinsic to the open-ended probe for low-loss materials make these measurements difficult. A very sensitive and accurate technique for determining low-loss sample properties is the *perturbation* technique. This measurement utilizes the change in frequency and the change in absorption characteristics of a tuned resonant cavity. The mea-

FIGURE 5–7.   Open-Ended Probe System for Measuring Dielectric Properties of Materials   (Courtesy Hewlett Packard Company).

surement is made by placing a sample completely through the center of a waveguide that has been made into a cavity. The cavity is made by placing two plates with central holes on either side of a section of waveguide, 1.5 guide wavelengths long. These plates, called irises, give the waveguide a very narrow frequency range for the transmission of microwave energy. Changes in the center frequency and width of this transmission characteristic, when a sample is inserted, provide information to calculate the dielectric constant and loss factor of the sample. For ease of measurement, a network analyzer can be used to automatically display changes in frequency and width (Engelder and Buffler 1991). A recommended waveguide cavity design with full theory and design details is available as a standard procedure published by the American Society for Testing and Materials (ASTM 1986).

Sample geometries can be of circular, rectangular, or square cross section with a recommended maximum dimension of approximately 0.125 in (0.318 cm) for a 2450-MHz system. Circular cross section is recommended because the measurement is independent of rotational orientation of the sample. This geometry however may be more difficult than a square cross section to fabricate. For a WR-284 cavity, a cavity length of 4.55 in (11.56 cm) is appropriate to give an approximate resonance frequency of 2450 MHz. A sample length of 2.5 in (6.35 cm) will adequately extend through the waveguide.

For solid materials, samples in the form of rods can be formed, molded, or machined directly from the material. For intermediate materials or liquids, samples may be filled or inserted into microwave transparent test tubes or tubing. Quartz is preferable, borosilicate glass is acceptable, but ordinary glass is not recommended. Tubing and test tubes can be purchased from chemical supply houses. Custom sample holders can also be obtained from a glassblower. Wall thickness should be as thin as practical, commensurate with required mechanical rigidity. Paper or plastic straws may also be used if glass is not available.

Liquids can be filled into test-tube sample holders with a pipet. Small-diameter pipets themselves also make excellent sample holders; use 200 microliters (μL) for low-loss materials and 10 μL for high-loss materials. Materials that can be melted can be poured into sample holders and allowed to solidify. This technique is appropriate if the material does not change its properties following melting and resolidification.

The ASTM standard provides details and formulas for calculating the dielectric constant and loss factor from the cavity parameters.

References

ASTM 1986. *Standard Methods of Test for Complex Permittivity (Dielectric Constant) of Solid Electrical Insulating Materials at Microwave Frequencies and Temperatures to 1650°C.* Document D 2520–86 (reapproved 1990). Philadelphia: American Society for Testing and Materials.

Ayappa, K., Davis, H., Crapiste, G., Davis, E., and Gordon, J. 1991. Microwave heating: An evaluation of power formulations. *Chemical Engineering Science* **46**(4):1005–1016.

Bengtsson, N. and Risman, P. 1971. Dielectric properties of foods at 3 GHz as determined by a cavity perturbation technique. Measurement on food materials. *Journal of Microwave Power* **6**(2):107–123.

Buckley, F. and Maryott, A. 1958. *Tables of Dielectric Dispersion Data for Pure Liquids and Dilute Solutions.* Washington, DC: National Bureau of Standards Circular 589. (Available through National Technical Information Service, Springfield, VA)

Buffler, C. and Stamford, M. 1991. The effects of dielectric and thermal properties on the microwave heating of foods. *Microwave World* **12**(4):15–23.

Debye, P. 1929. *Polar Molecules.* New York: Reinhold (reprint, 1945, Dover, New York).

El-Rays, M. and Ulaby, F. 1987. *Microwave Dielectric Behavior of Vegetation Material.* Radiation Laboratory, University of Michigan, Ann Arbor, MI. (Available through National Technical Information Service, Springfield, VA).

Engelder, D. and Buffler, C. 1991. Measuring dielectric properties of food products at microwave frequencies. *Microwave World* **12**(2):6–15.

Fricke, H. 1955. The complex conductivity of a suspension of stratified particles of spherical or cylindrical form. *Journal of Physical Chemistry* **59**:168–170.

Geyer, R. 1990. *Dielectric Characterization and Reference Materials.* NIST Technical Note 1338. National Institute of Standards and Technology, Boulder, CO.

Hewlett Packard 1985. *Measuring Dielectric Constant with the H/P 8510 Network Analyzer.* Product Note 8510-3. Hewlett Packard Corp., Palo Alto, CA.

Hewlett Packard 1991. Dielectric Probe Kit 85070A. Hewlett Packard Corp., Palo Alto, CA.

Kent, M. 1987. *Electrical and Dielectric Properties of Food Materials.* Hornchurch, UK: Science and Technology Publishers.

Marcuvitz, N. 1986. *Waveguide Handbook.* London: Peter Peregrinus. (Reissued on behalf of Institute of Electrical Engineers; available through IEEE Service Center, Piscataway, NJ).

Maxwell, J. 1873. *Treatise on Electricity and Magnetism.* Oxford: Clarendon Press.

Metaxas, A. 1985. A unified approach to the teaching of electromagnetic heating of industrial materials. *IJEEE* **22**:108–118.

Mudgett, R. 1985. Dielectric properties of foods. In *Microwaves in the Food Processing Industry* (R. Decareau, ed.), pp. 15–37. New York: Academic Press.

Mudgett, R. 1990. Developments in microwave food processing. In *Biotechnology and Food Process Engineering* (H. Schwartzberg and M. Rao, eds.), pp. 359–404. New York: Marcel Dekker.

Nelson, S., Stetson, L., and Schlaphoff, C. 1974. A general computer program for the precise calculation of dielectric properties from short circuited waveguide measurements. *IEEE Transactions on Instrumentation and Measurement* **23**(4):455–460.

Nelson, S. 1991. Dielectric properties of agricultural products—measurements and applications. *IEEE Transactions on Electrical Insulation* **25**(5):845–869.

Ohlsson, T. and Bengtsson, N. 1975. Dielectric food data for microwave sterilization processing. *Journal of Microwave Power* **10**(1):93–108.

Pace, W., Westphal, W., and Goldblith, S. 1968. Dielectric properties of common cooking oils. *Journal of Food Science* **33**:30–36.

Roberts, S. and von Hippel, A. 1946. A new method for measuring dielectric constant and loss in the range of centimeter waves. *Journal of Applied Physics* **17**:610.

Roebuck, B., Goldblith, S., and Westphal, W. 1972. Dielectric properties of carbohydrate-water mixtures at microwave frequencies. *Journal of Food Science* **37**:199–204.

Stuchley, M. and Stuchley, S. 1980. Coaxial line reflection methods for measuring dielectric properties of biological substances at radio and microwave frequencies—a review. *IEEE Transactions on Instrumentation and Measurement* **29**(3):176–183.

Tinga, W. and Nelson, S. 1973. Dielectric properties of materials for microwave processing—tabulated. *Journal of Microwave Power* **8**(1):23–65

von Hippel, A. 1954a. *Tables of Dielectric Materials. Dielectric Materials and Applications.* Cambridge, MA: MIT Technology Press. (out of print)

von Hippel, A. (ed.) 1954b. *Dielectric Materials and Applications.* Cambridge, MA: MIT Technology Press. (out of print)

von Hippel, A. 1954c. *Dielectrics and Waves.* New York: Wiley. (out of print)

6

Microwave Heating of Foods

INFLUENCE OF OVEN
AND PRODUCT PARAMETERS

Research over the past decade has determined that two major effects influence the heating patterns in microwaved foods (Buffler and Stanford 1991).
First, under certain circumstances, the hot- and cold-spot phenomenon discussed in Chapter 2 dominates the heating pattern. Under other circumstances the dielectric properties of the foods along with its geometry
dominate. In most circumstances there is a combination of geometry and
oven dependence on the heating pattern. It is well known that all microwave
ovens exhibit considerable variation of electric field inside the oven cavity
(Stanford 1990). Low-cost ovens, in general, have more hot and cold spots
than more expensive models, which usually have a well-designed stirrer or
a turntable. Oven domination of food heating pattern, in general, occurs
for small loads, particularly those with salt contents below 1% by weight.
Large loads of about 18 oz (500 g) and medium loads of about 11 oz
(300 g) containing 1% salt or higher tend to have heating characteristics
dominated by dielectric properties and geometry.

Effect of Penetration Depth

Chapter 5 has shown how the two microwave electrical characteristics, the
dielectric constant ϵ' and loss factor ϵ'', play a critical role in how microwave energy is deposited in a sample. The penetration depth, d_p of a microwave signal into a food product is determined by these two parameters, as
illustrated by the "food map" (Figure 5-1).

 The key to the use of the food map is to realize that all foods with the

69

same geometry and similar penetration depths will deposit microwave energy similarly. As shown, most foods tend to have penetration depths between 0.75 cm (0.3 in) and 1.0 cm (0.4 in). The actual temperature profile or heating pattern in the food is determined first by how the energy is deposited, as determined by its dielectric properties. Second, how the deposited energy is transferred throughout the sample is governed by the food's thermal and physical properties as well as the environmental or cooling conditions.

Foods with Slab or Flat Geometries

It is well known that foods in slab geometry form are difficult to heat, particularly at the corners and edges. Microwave radiation in the oven cavity can be crudely thought of as impinging on the food from all directions. Under this assumption, corners are especially vulnerable to heating because they are exposed to microwaves coming from numerous directions. Corners thus tend to heat more than edges, and centers tend to be quite cold. This phenomenon is illustrated in Figure 6-1. The shorter the penetration depth, which generally occurs with more salt, the more pronounced the corner and edge heating effects are. It is well documented that food products heat more uniformly if packaged in geometry with rounded corners. Oval or circular shapes reduce corner heating, with an optimal geometry being an annulus or doughnut configuration.

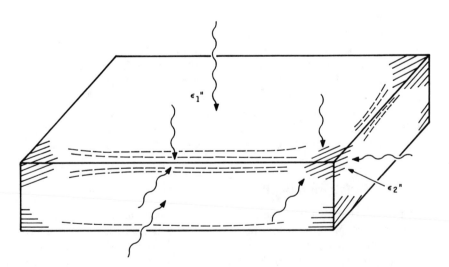

FIGURE 6-1.   Edge Heating Effects in Slab Geometry   (Buffler and Stanford 1991).

Focusing in Spherical
and Cylindrical Geometries

An interesting phenomenon occurs with spherical and cylindrical geometries that does not occur with flat, rectilinear geometries, namely focusing. Consider what happens to random microwave signals impinging on a cylindrical or spherical food product in a microwave oven. For high-dielectric materials, such as the foods illustrated on the food map, it can be demonstrated that the microwaves will be "aimed" or refracted toward the center. The aiming phenomenon comes about due to Snell's law (Equation 5-4), which relates the angle of incidence, α, of a wave in air ($\epsilon' = 1$) to the angle of refraction (the angle inside a dielectric medium), β (cf. Figure 6-2). Calculating from Snell's law, for a dielectric constant $\epsilon' > 40$, and using a worst-case incident angle of 90°, one finds that all incident rays on the food are refracted to within 9° of the internal normal.

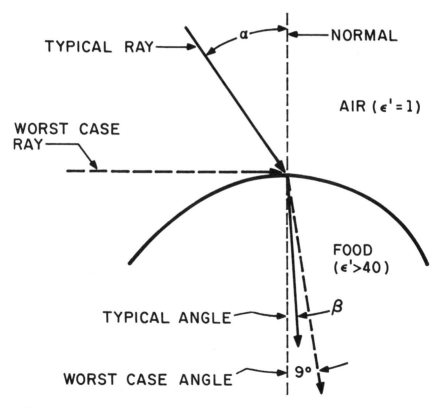

FIGURE 6-2.   Microwave Refraction at Food Surface   (Buffler and Stanford 1991).

The focusing effect due to these refracted rays depends directly on the penetration depth and can be simplistically described: If the penetration depth is large compared with the dimensions of the food, most of the waves progress through the product without being greatly attenuated. If the penetration depth is quite small compared with the dimension of the food product, most of the energy is absorbed within one penetration depth from the food surface, causing intense heating at the surface but leaving the center cold (Figure 6-3a). For intermediate penetration depths, the microwave energy is not completely absorbed in the outside layer, and a reasonable amount reaches the center. Because the microwave rays are entering the product through all surfaces, they all become concentrated in the central region before being greatly attenuated. Thus, concentrated heating, or focusing at the center, occurs (Figure 6-3b). This focusing effect has been demonstrated theoretically (Ohlsson and Risman 1978), as well as being verified by the consumer! It is the cause of center burning of overmicrowaved potatoes and the "bumping" or "blurping" effects seen with soups and stews in cylindrical containers. Focusing also contributes to the eruption phenomenon that occasionally occurs when water, coffee, or tea is heated (Buffler and Lindstrom 1988).

To illustrate focusing, a crude mathematical model can be evoked. If one assumes all microwave energy enters a food radially and neglects reflection from internal boundaries, one can calculate the power dissipated in concentric shells as the microwave energy propagates toward the center. With this model it is easily seen that maximum focusing occurs when the penetration depth is approximately 1.5 times the food sphere diameter. Calculations for a cylinder yield similar values. From these calculations and measurements it can be seen that the approximate optimal diameter for uniform heating

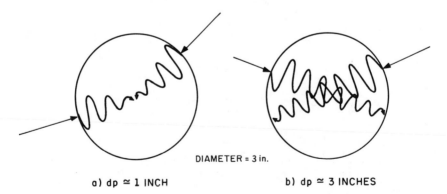

DIAMETER = 3 in.

a) dp ≃ 1 INCH   b) dp ≃ 3 INCHES

FIGURE 6-3.   Focusing Effects in Spheres and Cylinders   (Buffler and Stanford 1991).

is approximately two to two and one-half times the value of the penetration depth in the food. This rule of thumb should always be considered in the development of product geometries.

THERMAL PROPERTIES OF FOODS

Two thermal properties, thermal conductivity k (W/m K) and heat capacity (often erroneously called specific heat; cf. glossary), C_p (J/kg K), and two mechanical properties, density, ρ (kg/m³) and viscosity, η (Pa-s), determine how a food product heats after microwave energy has been deposited in it. Thermal properties of foods have been tabulated in the literature (Morley 1972; Polly, Snyder, and Kotnour 1980), with some being listed in Appendix 3. If heat capacity thermal conductivity, and density are not known, it is quite simple to determine them with good accuracy if the proximate analysis is known. For a large list of foods, proximate analyses can be found in *Agricultural Handbook 8* (1963). Empirical equations for the above parameters have been developed by Choi and Okos (1986) and are for dense foods containing little air at 20°C (68°F). The values have been rounded:

$$C_p \text{ (J/kg K)} = 4190W + 1780P + 1980F + 1420C + 950A \qquad \text{(6-2a)}$$

$$k \text{ (W/m K)} = 0.6W + 0.2P + 0.18F + 0.2C + 0.14A \qquad \text{(6-2b)}$$

$$\rho \text{ (kg/m}^3\text{)} = 1000W + 1290P + 920F + 1430C + 1740A \qquad \text{(6-2c)}$$

Where W, P, F, C, and A are the volume fractions of water, protein, fat, carbohydrates and ash (W + P + F + C + A = 1).

Below 0°C (32°F) the thermal and physical properties of the food depend directly on the amount of water that remains unfrozen at a given temperature. The energy required to raise a product's temperature to a given level is given by its enthalpy curve. Three products are illustrated in Figure 6–4 (Tressler, van Arsdel, and Copley 1968). The inverse slope of the curve gives the effective heat capacity of the material at a given temperature. Below −10°F, all three products have a heat capacity of 0.6 Btu/lb°F (2514 J/kg K), where all water possible has been frozen. Above 32°F (0°C) the heat capacities are somewhat different. Between −10°F (−23°C) and 32°F (0°C) the effective heat capacities are very large due to the latent heat of fusion required to melt the frozen water. For many food products, some

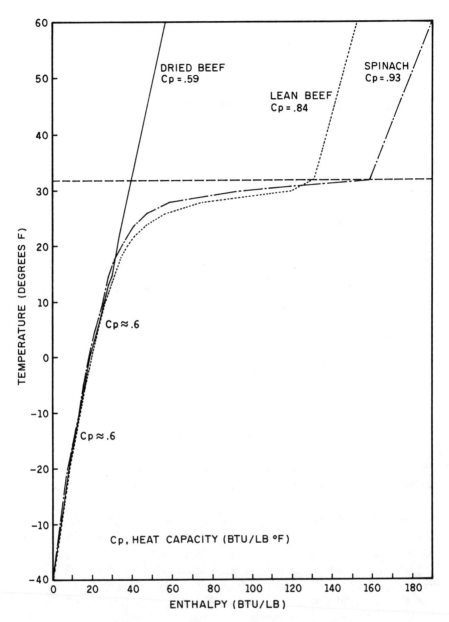

FIGURE 6-4.   Enthalpy Curves for Meat and Vegetables   (From data in Tressler, van Arsdel, and Copley 1968).

portion of the water in the food melts over the entire temperature range from $-10°F$ ($-23°C$) to $32°F$ ($0°C$).

The thermal heat capacity, C_p, determines how rapidly a food heats once the power is deposited within it. The rate of temperature rise within a food is

$$\frac{\Delta T \ (°C)}{\Delta t \ (\text{min})} = \frac{P \ (W)}{14.7 \times V \ (\text{m}^3) \times C_p \ (\text{J/kg K}) \times \rho \ (\text{kg/m}^3)} \qquad (6\text{-}3a)$$

or in calories, grams, and liters,

$$= \frac{P \ (W)}{70 \ V \ (\text{L}) \ C_p \ (\text{cal/g °C}) \ \rho \ (\text{g/cm}^3)} \qquad (6\text{-}3b)$$

Since the power absorbed depends on the dielectric properties and the heating rate on the thermal properties, in a multicomponent food product, where the components have widely differing dielectric and thermal properties, it is often necessary to balance both sets of properties in order to approach equal heating for each component. It is usually more fruitful to adjust specific heat, if possible, rather than dielectric properties to obtain such a balance.

Viscosity determines how a product will flow or convect when heated. The mathematics for predicting heating in a convecting medium are extremely complex and have not been sufficiently explored in the literature. In predicting heating of convective materials, one can at best examine what rules of thumb apply as if it were a solid, realizing that errors may become extreme when viscosity is low.

The one-dimensional differential equation that describes the temperature rise with time of a material is

$$\frac{\delta T}{\delta t} = \alpha \frac{\delta^2 T}{\delta x^2} \qquad (6\text{-}4)$$

where the thermal diffusivity is defined as

$$\alpha = \frac{k}{C_p \rho} \qquad (6\text{-}5)$$

the ratio of conductivity to the product of heat capacity and density. For x in meters and t in seconds, the units of thermal diffusivity are m^2/s. Equation 6-4, and thus α, determines how rapidly heat moves or diffuses through a product. Figure 6–5 illustrates the thermal conductivity plotted against the product of heat capacity and density for numerous foods. The lines of constant slope represent lines of equal thermal diffusivity as given by Equation 6–5. The material with the highest diffusivity is pure frozen water or ice. Thus, the more frozen water a food contains, the faster heat will move away from the point where energy is deposited.

A review of Figure 6–5 as well as the literature, curiously revealed that virtually all foods of interest to microwave product developers have thermal diffusivities within a narrow range of values on either side of 1×10^{-7} m^2/s, if they are in the thawed state. If mass transfer effects, which take place at temperatures near boiling, are neglected, these foods will heat with similar thermal profiles if microwave energy is initially deposited within them in an identical manner.

Differences in environmental or cooling conditions influence the temperature distribution near the product surface. Examining the magnitude of airflow in microwave ovens, one finds that the heat transfer conditions have minimal effect on the heating pattern of a nonenclosed item. The heating rate and pattern, however, of foods heated in packages or containers depends critically on the insulative properties of the package configuration!

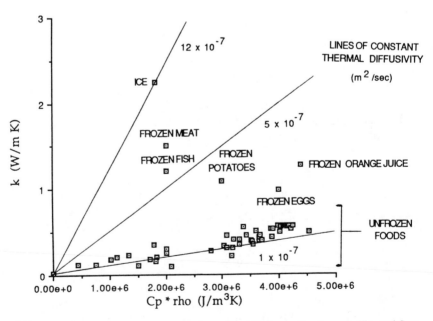

FIGURE 6–5.   Thermal Properties of Thermal and Unfrozen Foods   (Buffler and Stanford 1991).

Frozen Foods

The heating or thawing of frozen foods basically follows the rules of thumb for thawed foods; that is, they follow the dependence on oven parameters, geometry, and penetration depth. The dielectric parameters of frozen food are complicated because they depend not only on the amount of unfrozen water, which changes with temperature, but also on its location and salinity. Penetration depths therefore range typically from 10 cm (4 in) for deeply frozen foods to below 1 cm (0.4 in) near 0°C (32°F). The penetration depth for pure water (ice) at 0°C (32°F) is 1162 cm (38 ft)!

The most severe complication of heating or thawing frozen foods comes from the geometry consideration previously discussed. Corners and edges, exposed to concentrated microwave energy, thaw first (Figure 6-1). These thawed areas now have much higher dielectric loss, ϵ_2'', and thus heat very rapidly at the expense of the still frozen regions, ϵ_1''. This phenomenon, known as *thermal runaway,* is the reason that the defrost cycle is set at approximately 300 W on most microwave ovens. The low power setting allows time for the heat generated on the corners and edges to diffuse into the frozen interior.

MICROWAVE HEATING "RULES OF THUMB"

Rules of thumb and practical considerations for predicting microwave heating patterns and developing microwavable food products are summarized from the preceding sections:

1. Calculate penetration depths as a function of ϵ' and ϵ'' for food products and food systems of interest. Plot on the food map.
2. Most practical food products fall in similar regions on the food map and thus have similar penetration depths.
3. Predict whether dielectric/geometry characteristics or oven characteristics dominate the heating pattern. Determine experimentally by testing in several microwave ovens.
4. For most products, penetration depth and edge effects are the most important parameters in determining the heating pattern. Review possible geometries with respect to corner and edge heating and focusing. Adjust geometry where possible.
5. In cylinders and spheres focusing occurs when penetration depth is approximately 1.5 times their diameter. Adjust penetration depth where possible, with formulation, assuming thermal diffusivity remains approximately constant.
6. For multicomponent products with differing heating rates and different

dielectric and thermal properties, consider adjusting the specific heat before adjusting the dielectric properties. Heating rate is inversely proportional to specific heat.

7. Thermal diffusivities of *un*frozen foods are similar. Foods heat similarly for equivalent energy deposition.

8. Frozen foods are very sensitive to thermal runaway. Use the above criteria at a lower power if possible.

There is no question that the development of microwavable food products is difficult and frustrating. Using the above considerations, however, along with a solid foundation of food science, the speed and success of the development should be increased. Patience as well as trial and error is also a requisite.

COOLING OF MICROWAVED FOODS

The cooling of microwaved foods can be divided into two categories: evaporative cooling during microwave preparation, and cooling after foods have been heated in a microwave oven.

Microwave heating and cooking are unique in that the oven and surrounding air are cooler than the product as it cooks, in contrast to preparation in a conventional oven. Under these circumstances the maximum temperature that water-containing foods reach is the boiling point of water. This temperature depends on the atmospheric pressure at a particular location (cf. Table 10-3). The boiling temperature is maintained until the product is desiccated, at which point the temperature may or may not rise, depending on the dielectric loss characteristics of the dried material. Dried materials with high loss, such as those containing cellulose, may produce thermal runaway, resulting in fires (Chapter 9). Materials with low loss have a self-limiting effect since they no longer heat when dried. This phenomenon is valuable for industrial processes (Chapter 10).

During the heating process of foods containing water, the resulting evaporation at the surface causes a depression of the temperature, known as evaporative cooling. This phenomenon is readily seen during the cooking of a meat roast. Figure 6-6 illustrates a roast cooked in a microwave oven after various cooking times. The initial temperature of the roast upon removal from the refrigerator is uniform throughout at 10°C (50°F). Because the microwave penetration depth of the meat is only 1 cm (0.4 in), heating is at the surface; the interior temperature rise being primarily due to heat conduction. As cooking proceeds, the surface temperature approaches 100°C (212°F) and moisture evaporation occurs. The evaporation depresses the surface temperature while the temperature just inside the surface may

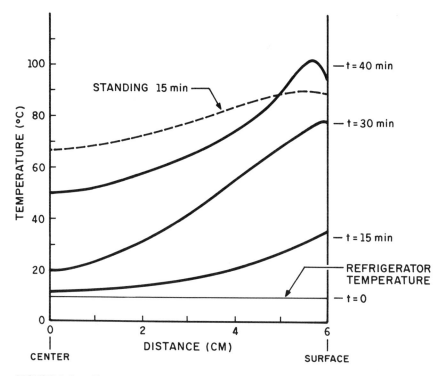

FIGURE 6-6.   Temperature Profiles in a Roast during Microwave Cooking.

be at 100°C (212°F). The temperature difference from outside to inside, or the *temperature gradient,* is considerably higher than that found in a conventionally prepared roast.

If the roast is allowed to stand, preferably covered with foil following cooking, the internal temperature thermalizes or redistributes. Heat from the outer portion of the roast moves inward by conduction; therefore, microwave-prepared meats should be removed from the oven before the center temperature reaches the desired doneness. The meat should be allowed to undergo a standing time to thermalize to the desired temperature before serving.

Since the advent of microwave cooking there have been apocryphal stories relating to the rapid cooling of foods prepared in a microwave oven compared to conventional techniques. Mysterious properties of microwaves were imputed to be the cause of this rapid heat loss. There are no magical properties of microwaves! Cooling of microwaved food obeys the same laws of physics as those conventionally prepared, so if there are differences in cooling rates there must be a rational reason. The subject is extremely complex, with each example of food preparation having to be considered separately. Specifically, if there is any difference in cooling rate between the two processes, it must arise from a difference in initial internal temperature of the food or from differences in conductive or convective heat loss of the serving environment. As an example, probably the most common reason that microwaved dishes cool more rapidly arises from the fact that the food surface and the serving dish remains relatively cool during microwave cooking. When the dish is brought to the table, heat is transferred to the dish, cooling the food. Food prepared in a conventional oven, on the other hand, has a hotter surface temperature and is brought to the table in a hot dish, which keeps the food warm. A detailed discussion of the various cooling mechanisms is given by Buffler (1991).

MATHEMATICAL MODELING

As indicated in Chapter 5, if an exact knowledge of the microwave oven cavity configuration, food or product geometry, and dielectric properties is available, one can use Maxwell's equations to calculate the exact microwave electric field configuration within the food. With this knowledge, along with the physical, thermal properties of the food, and the environmental cooling conditions, the heating pattern of the food can be determined. Unfortunately, an analytical or exact solution of Maxwell's equations is extremely difficult or impossible for all but the simplest product geometries and cavity configurations. Thus, to make analytical calculations of field configurations and heating patterns, one must make simplifying assumptions that may greatly compromise the usefulness of the results obtained.

An alternative is to use sophisticated mathematical procedures, known as modeling. This technique breaks down the cavity and food geometries into small regions or cells. Maxwell's equations can be approximated and solved for each cell. The individual cell solutions are combined to form a final solution for the electric field within the entire configuration. The electric field defines the deposition of energy within the sample according to Equation 5-3. If the thermal properties and the external cooling conditions of the product are known, the power deposited can be used to determine the temperature rise pattern in the sample. This calculation is done by using heat transfer equations that have also been broken up into the same cells used for Maxwell's equations. The heat transfer equations are similar to the one-dimensional Equation 6-4, but are three-dimensional and much more complicated. Fortunately these heat transfer equations have been computer-programmed and are readily available as commercial software (Lorenson 1990).

Modeling techniques can provide excellent results if the cell size is sufficiently small, usually less than one sixth of the wavelength within the region of interest. Unfortunately, as cell size diminishes, computer processing time increases, requiring larger, high-speed computers to obtain accurate results. Some saving in computer time can result from judicious consideration of symmetry. For example, for a product in the center of an oven shelf in a cavity with top feed, only one quadrant of the cavity need be modeled; the other three quadrants have the same heating pattern by symmetry.

Modeling techniques require extensive computer programs and are beyond the scope of this text. Many researchers working in microwave technology, however, at one time or another will find it advantageous to model their product and thus need to interact with consultants or firms providing such services. For researchers to have intelligent discussions with these resources, a brief review of the three most commonly used modeling techniques is presented with advantages and disadvantages of each. A comprehensive and detailed review of the different techniques has been given by Lorenson (1990).

Finite-Difference, Time-Domain (FD-TD) Method

For the FD-TD the region to be studied is divided into cells, usually cubic. Each cell is assigned the value of the complex dielectric constant ϵ at its location. The mathematics is easier if a food with uniform ϵ is assumed. Maxwell's equations are approximated by assuming that there is a linear relationship between the fields on each side of the cell. The incident wave is followed as it propagates into the sample, and new electric fields are calculated along the way. By this means, the electric field buildup can be observed (transient state) as well as the final field configuration (steady state). The advantage of the FD-TD approach is the simplicity of mathematics required to describe the mesh of cells. For this reason it is the most frequently used technique for microwave modeling. The disadvantage of the technique is the long computer processing time required, because all space must be broken up into cells of regular shape.

Finite-Element Method (FEM)

Finite elements are equivalent in principle, to the cells of the FD-TD method, but can be any configuration that best conforms to the geometry to be modeled. FEM allows different-sized elements to be used for different regions. For example, large elements might be used in the cavity where the wavelength is long. Small elements can be used inside the food where the wavelength is short and better pattern resolution is desired. A major advantage to FEM is that the power deposition solutions can be performed first and then transferred to a separate computer program where the temperature profile can be calculated. A major difficulty of the technique is the difficulty of treating the boundary between air and food. Skill and experience are needed to obtain valid solutions in this boundary region.

Method of Moments

The method of moments is most powerful when plane-wave illumination can be assumed. Techniques have been developed to take into account the multiple reflections that occur inside an oven cavity so that adequate solutions for microwave oven configurations can be obtained. This model divides the region of interest into a mesh or cells that differ radically from those of the FD-TD or FEM techniques. Here the edges of the cells are represented by fine wires, and attached to them are lossy inductors or capacitors (lossy elements) that simulate the food. The exciting microwave field is represented by generators attached to "appropriate" portions of the wire mesh and provide known amounts of current to the wires. The electric field inside each cell is then calculated from the current flowing in the wires. The advantage to this method arises from its mathematical simplicity as well as its ability to model metal boundaries. These metallic boundaries can easily be modeled with wires of various conductivities. The interaction of susceptors, shields, and metal foil structures in packages with the food product can thus be successfully modeled. A disadvantage is that we cannot know exactly where and how to connect the exciting generators to simulate the oven feed conditions. It is also difficult to perform calculations on wire configurations with nonrectangular shapes. This modeling technique also requires skill and experience.

References
Agricultural Handbook 8. 1963. Washington DC: U.S. Department of Agriculture. (Reprinted in 1975 as *Handbook of the Nutritional Contents of Foods*. New York: Dover).
Buffler, C. 1991. Perceived rapid cooling of microwaved foods. *Microwave World* **12**(2):16–18.
Buffler, C. and Stanford, M. 1991. The effects of dielectric and thermal properties on the microwave heating of foods. *Microwave World* **12**(4):15–23.
Buffler, C and Lindstrom, T. 1988. Experimental evidence of water eruption caused by super-heating. *Microwave World* **9**(4):10–11.
Choi, Y. and Okos, M. 1986. Thermal properties of liquid foods—review. In *Physical and Chemical Properties of Foods* (M. Okos, ed.), pp. 35–57. St. Joseph, MI: American Society of Agricultural Engineers.

Lorenson, C. 1990. The why's and how's of mathematical modelling for microwave heating. *Microwave World* **11**(1):14–23.

Morley, M. 1972. *Thermal Properties of Meats: Tabulated Data.* Langford, Bristol, UK, BS18 7DY: Meat Research Institute.

Ohlsson, T. and Risman, P. 1978. Temperature distribution of microwave heating— spheres and cylinders. *Journal of Microwave Power* **13**(4):303–310.

Polly, S., Snyder, O., and Kotnour, P. 1980. A compilation of thermal properties of foods. *Food Technology* November 1980.

Stanford, M. 1990. Oven characterization techniques and how to use the data. Proceedings of the 26th Annual Symposium of the International Microwave Power Institute, Buffalo, New York.

Tressler, D., van Arsdel, W., and Copley, M. 1968. *The Freezing Preservation of Foods.* Vol. 2. *Factors Affecting Quality in Frozen Foods.* Westport, CT: AVI.

7

Packaging, Containers, and Susceptors

Much literature is available on the topic of packaging for food. Basic texts (Bakker 1986; Sacharow and Brody 1986), handbooks (Hanlon 1984), trade journals (*Packaging*), and directories (Cahners 1989, 1991) should be studied for those pursuing careers in packaging technology. This chapter will not focus on general fundamentals but will concentrate on those aspects of packaging that directly relate to microwave cooking and heating, emphasizing microwave *active packaging.* An active package is one that changes the electric (or magnetic) field configuration and thus the heating pattern of the product contained within. Active packaging also includes susceptors or heater boards that are included in a package to brown or crisp a product. Except for special susceptors, magnetic materials are not generally used as microwave packaging materials and will not be treated in this book.

Passive packaging, on the other hand, is packaging that does not appreciably effect the microwave heating pattern; that is, it is essentially transparent to the passage of microwaves and is covered in the references.

REFLECTION, ABSORPTION, AND TRANSMISSION

Transparent Packaging Materials

As described in Chapter 5, all materials exhibit properties of reflection, absorption, and transmission when exposed to electromagnetic radiation. Figure 5-3 depicts the behavior of an electromagnetic or microwave ray impinging on the surface of a material. This material may be quite thick or, more commonly, with packaging materials, quite thin. Microwave packaging materials are usually desired to be transparent to the microwaves im-

84

pinging on them. Effective transparency is easily possible with materials chosen to have a low dielectric loss factor. Thin materials, with low losses, exhibit a fortunate microwave property in that they do not introduce appreciable reflection at the air-dielectric boundary even though the dielectric constant ϵ of the material is appreciable. This phenomenon arises because the reflection from the initial surface that the microwaves encounter (for example, the reflection given by Equation 5-2) is almost completely canceled by the internal reflection from the opposite surface (cf. Figure 5-3). The electric field of the wave reflected from the opposite surface is equal and opposite in direction to the wave that reflects off the first surface. This difference in field direction occurs because the initial reflection is due to an air-dielectric interface while the second surface reflection is due to a dielectric-air interface. Since the material is of low loss, there will be minimal attenuation when the wave traverses the thickness d of the material and returns. If d is very small compared with the wavelength in the material, the wave essentially has not changed by the time it returns to the first surface. The exiting wave thus almost entirely cancels the first reflected wave. With virtually no net reflection nor absorption in the material, practically all the microwave energy is transmitted through the thin sheet of material, since $P_t = 1 - P_r - P_a$.

Shielding

As previously discussed, highly conductive metal in sheet or foil form virtually reflects all microwave energy impinging on it. This property can be extremely useful if microwave energy is desired to be excluded or limited from an internal portion of the package. A rule of thumb for most food products with penetration depths around 1–2 cm (0.4–0.8 in) is that the amount of microwave energy entering a product is directly proportional to the area of the product exposed to microwave radiation. Thus, a completely enclosed metal container allows no microwaves to enter; a shallow metal dish or tray with an open top, on the other hand, allows approximately 50%. Foil is usually used as a shielding material in package design and is especially useful for multicomponent meals where the different portions require different heating rates (Moffett 1952; Welch 1955). For some products it may be desirable to shield the majority of the package and allow only a small portion of microwave energy to enter. This effect can be accomplished by providing holes in an otherwise completely shielded package. Holes have a unique property in that they allow microwaves to propagate through them when they are larger than a critical dimension, and completely "cut them off" or exclude them when they are smaller. The critical dimension depends on the hole geometry, hole pattern, metal or foil thickness, and free-space wavelength λ_0. The theory of the cutoff phenomenon is complex and can be found in most texts on electromagnetic theory (Marcuvitz 1986; Harrington 1987). It is sufficient to remember that the cutoff dimension for a square hole of dimension a is approximately $\lambda_0/2$. For a circular hole, the cutoff diameter b is approximately $\lambda_0/1.71$. For 2450 MHz these dimensions are $a = 6.12$ cm (2.41 in) for a square hole and $b = 7.15$ cm (3.82 in) for a round one. The equations for cutoff dimensions theoretically hold only for long tubes of the appropriate cross section. Calculations for various arrays of holes in sheets with finite thickness have been calculated by Otoshi (1972) and Chen (1973). This cutoff phenomenon allows microwave oven door screens to be used to view the food during preparation. The wavelength for light is so small that the waves easily propagate through; the microwaves, however, are cut off because of their large wavelength and cannot escape.

For prototype package fabrication it is usually not practical to perform mathematical design calculations. Basic understanding of the above considerations allows the designer to use empirical or cut-and-try techniques more effectively. When microwave package prototypes are prepared, it is extremely useful to fabricate from paperboard, plastic, and aluminum tape if shielding is to be considered. Aluminum tape, found at most hardware stores, is applied in the place of the desired shielding, cutting many hours from prototype testing and evaluation.

Field Modification Techniques

Metal foil in the form of strips or patches may be used to modify and intensify the microwave electric field. These patches act as small antennas to capture and then reradiate the impinging microwave energy onto the food product. Field distributions can be modified to compensate for nonuniform heating patterns or intensified to produce browning (Keefer 1987). Field modification can be accomplished with either metal or microwave transparent bowls, dishes, or trays. The topic of field modification has been reviewed by McCormick (1991).

PRACTICAL MICROWAVE PACKAGE DESIGN CONSIDERATIONS

A package must perform various functions and they are discussed individually. Comments are given relating to how the material choice must be made to balance the required function with the fact that the package is to be used in a microwave environment.

Advertisement, Communication, and Decoration

The exterior of a package is designed to communicate information and aid marketing by catching the consumer's eye. If any portion of the decoration or graphics is to remain on the package when it is placed in the microwave oven, several precautions must be observed.

1. Metal foil, in general, is not recommended for use on a microwavable package. Since foil is electrically conductive, it reflects microwaves and should not be used unless it has been designed to act as a shield or as an integral portion of an active package design. In addition, many decorative films utilize a thin layer of deposited metal to give them their optical reflective qualities. The metal particles in these films are deposited so thinly that there are microscopic spaces between them. In microwave applications arcing will occur between the particles, which will destroy the film and could produce a fire.

2. The inks used for graphics should be carefully specified as nonconductive. Some carbon-based inks or those with conductive particles may heat, char, or melt the substrate.

In all cases, it is best to perform microwave testing on prototype graphics before production is begun. Worst-case testing of the package without the product will give the highest level of performance confidence.

Protection

Packaged foods must be mechanically protected from damage during distribution, storage, and shelving. Also, environmental contamination from soil, dirt, and microorganisms must be prevented so that a safe and attractive product may be delivered to the consumer. For mechanical protection from damage, conventional packaging design criteria should be used for strength and rigidity. Selection of packaging materials should be based primarily on these criteria with attention paid to their microwave properties. In general, package materials should be transparent to microwaves and thus have a low dielectric loss factor, preferably less than 0.05. Fortunately, most plastics and paperboard are low-loss, but detailed dielectric properties are not readily available for these materials at microwave frequencies. A few materials are listed in Appendix 3.

Two notable exceptions to low-loss packaging materials exist:

1. Nylon is hygroscopic and may contain or absorb enough water to heat appreciably under microwave exposure. Nylon is not recommended as a microwave packaging material, but if used it should undergo extensive applications testing.

2. Uncoated paperboard and paper are hygroscopic and pick up moisture from the air. If a microwavable product is to be sold and/or stored in areas that have high relative humidity and no air conditioning, care should be exercised in using these materials. If such materials, containing too much moisture absorbed from the atmosphere, are placed in a microwave environment, they may overheat, desiccate, and char or burn.

It is best to request information, such as dielectric constant and loss factor at 2450 MHz, from material suppliers for their products. These parameters as a function of temperature and relative humidity are especially important. Most manufacturers cannot supply such information. Continued requests, however, may eventually influence vendors to make microwave measurements on their own materials.

Serving

Many microwave meals are designed to be eaten directly from their packages. When a container is chosen for such a product, the manufacturer must decide whether the product is to be prepared in the microwave oven only or whether it will have instructions for preparation in a microwave oven as well as a conventional oven. The former are typically called *microwave only* products; the latter, *dual ovenable.*

1. Dual-ovenable materials must be able to withstand the temperatures found in a conventional home oven, typically 400–450°F (204–232°C). Most conventional thermoplastics, which can economically be used for container materials, will generally not withstand these temperatures. Thermoset polyesters, and occasionally CPET and high-temperature polyester-coated paperboard are used instead. Dishes made from thermoformed crystallized polyester (CPET) may be used for conventional or microwave oven heating.

2. Microwave-only materials need not be as critical concerning high-temperature performance. Most microwave-only products contain a considerable amount of water. Thus, the temperature of the food and container usually does not exceed the boiling temperature of water. Materials such as crystallized polyester, filled or unfilled polypropylene, or high-density polystyrene may thus be used for packages for these foods.

A strong note of caution should be expressed for foods containing a high amount of fat. These foods may contain but little water and tend to desiccate severely around the edges when microwaved for longer than the recommended heating time. Under these extreme conditions the temperature of the remaining fatty food at the edges may exceed the deflection, and even melting, temperature of the plastic container. Potential for spillage thus increases, and the consumer could be burned when removing the product from the microwave oven. It is always wise to abuse test prototype products and packages to make certain that the combination is safe in microwave oven applications.

Sealing

Closing and sealing are the final step in protective packaging. Sealing protects against contamination from dirt and microorganisms. The seal, as well as the package itself, acts as a gas and moisture barrier from the external environment into the package as well as from the package to the outside. Sealing techniques are discussed in conventional packaging references. A few considerations, however, are unique to products to be prepared in the microwave oven.

Shelf-Stable Products

The fact that microwaves penetrate plastic containers offers a unique convenience for the microwave preparation of ambient-temperature, shelf-stable products that have been sterilized by retorting or hot filling. Technology has recently been developed whereby sterilization can be effected directly in a sealed plastic container instead of conventional steel or aluminum cans. This technology was developed as an extension of the retort pouch technology developed primarily for armed service use (Lampi 1977).

If a dual-ovenable container is required, crystallized polyester trays are recommended. This container material can be sealed and retorted with aluminum foil lidstock or with microwave transparent plastic lidstocks with layers or coatings of polyvinylidene chloride (PVDC), ethylene vinyl alcohol (EVOH), or polyester. All have been shown to provide the requisite microbiological, moisture, and oxygen barriers required for many shelf-stable products. The metal-versus-transparent lidstock choice should be evaluated for their comparative microwave processing performances. All approaches should be compared for cost and environmental acceptability. If a microwave-only product is acceptable, polypropylene trays may be used with microwave transparent lidstock as long as both tray and lid have sufficient barrier properties, such as layering or coating with PVDC or EVOH.

THEORY AND PRACTICE OF SUSCEPTORS

A major drawback of microwave food preparation is the lack of browning of foods. Lack of browning occurs because as the food heats and evaporates moisture from its surface, the surface does not become hotter than the water's boiling temperature. Temperatures of over 350°F (177°C) are required to produce effective browning. One way of overcoming this problem is by using an external means to provide a temperature high enough to effect browning. Early solutions to the problem were the use of browning or crisping dishes. Browning dishes arose from early experiments which were carried out by coating borosilicate pie plates with a thin layer of tin oxide. The concept was later commercialized by using custom-designed glass ceramic dishes. When exposed to the microwave electric field, the oxide layer on the dish absorbs energy and becomes extremely hot. By this means, the temperature of the dish could be raised to 400–500°F (204–260°C). Placing a food product, such as a chop or meat pattie, on a preheated dish produces a nicely seared and browned product. A major disadvantage of the browning dish was that the dish had to be reheated a second time to sear the opposite side. A "waffle iron"–type accessory was later developed to simultaneously sear both sides.

As microwave products became more prevalent, the need for browning without an external accessory became apparent. The use of a square of tin-oxide-coated glass ceramic placed in a package to crisp pizza was attempted at Pillsbury in the mid-1970s (Turpin 1989). From that concept, adaptations resulted providing an inexpensive, disposable structure that could be incorporated into each package.

Since a resistive film was already known to heat when deposited on a substrate, an inexpensive heater was conceived by first depositing a thin film of metal on a thin plastic film of polyester (Brastad 1980; Seiferth 1987). Initially aluminum was used with a layer thickness of about 10–20 angstroms. This aluminized film was then laminated to paperboard stock with an adhesive. The sandwich became known as a susceptor heater board or simply susceptor and is illustrated in the insert of Figure 7–1. One of the first commercial applications of the susceptor was in a frozen Pillsbury pizza package. The pizza, resting on a disk of susceptor board, was heated in the package in a microwave oven. The resulting browning and crisping of the underside produced a product close in quality to one prepared by conventional means. Today, susceptors are used in many microwavable products.

Several reviews of susceptors have appeared in the literature (Turpin 1989; Pesheck 1990).

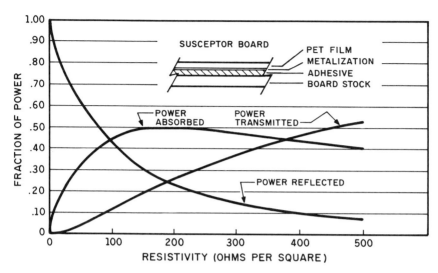

FIGURE 7-1.   Reflection, Absorption, and Transmission vs. Resistivity of a Susceptor Film (Buffler 1991).

Theory of Operation

It has long been known qualitatively that a metallic sheet of very low resistivity (high conductivity) reflects virtually all microwaves impinging on it. Virtually no energy is absorbed in the sheet, since all has been reflected and none is available. Conversely, a sheet with extremely high resistivity (low conductivity) is essentially an insulator and practically all microwave energy propagates through it without being absorbed. Thus, for zero and infinite resistivity, a metallized sheet has zero power absorption and thus no heating. If the sheet does absorb energy at some value of resistivity, it must reach at least one maximum of absorption between these two end-point values. Qualitative curves to this effect have been published, but values are not presented. In earlier works (Ramey and Lewis 1968; Ramey et al 1968), Maxwell's equations were solved to give quantitative results, but the mathematics is arcane to food scientists.

To calculate the absorption of a susceptor as a function of resistivity, an electronic equivalent circuit can be developed and simple equations derived for the reflection and absorption of incident power (Buffler 1991). The equations are based on the discontinuity that the film's resistivity produces compared with the "resistivity" of free space. The following equations for fractional reflected power, P_r, absorbed power P_a, and transmitted power

P_t may be easily incorporated into a spreadsheeet program to calculate fractional powers as a function of film resistivity:

$$P_r = \frac{1}{(1 + 2r)^2} \tag{7-1}$$

$$P_a = \frac{r}{(r + 0.5)^2} \tag{7-2}$$

where $r = R/Z_0$, where R is the film resistivity in ohms/square* (Ω/sq), and Z_0 is the resistivity of free space, equal to 377 Ω. The fractional transmitted power, as discussed in Chapter 5, is

$$P_t = 1 - P_r - P_a \tag{7-3}$$

Equations 7-1, 7-2, and 7-3 are plotted in Figure 7-1. The maximum value of absorption, 50%, occurs at a film resistivity value of $Z_0/2 = 188.5$ Ω/sq. The theoretical values agree well with experimental results (Lindstrom 1990) and qualitative curves (Andreasen 1988; Turpin 1989; Turpin and Perry 1990). Further work should be performed to verify heat generation as a function of resistivity by a heat flux technique described by Huang (1990).

Practically, it is difficult to measure susceptor resistivity by using metal electrodes, because of resistance errors introduced at the electrode-susceptor interface. Commercial, noncontacting, resistivity-measuring equipment is presently commercially available (Blew 1989), which avoids this problem.

Practical Susceptors

The most severe problem with thin-film susceptors, if laminated to paperboard, is ignition. If the paperboard temperature is raised to over 451°F (233°C), charring and burning can occur. The problem is partially solved in two ways. First, the laminated susceptor is designed so that the continu-

---

*The unit of film resistivity, ohms/square, is commonly used for specifying thin resistive sheets. The units represent the resistance, measured in ohms, of a square cut from the film material. The measurement theoretically is made by placing two electrodes along opposite sides of a film square; thus, the electrode length is equal to the distance between them. The measured resistance is *independent* of the dimension of the square, as can be seen from the following details: If a resistance is measured with a given electrode length and a spacing equal to that length, and the electrode length is doubled, the resistance decreases to half its value. This procedure is equivalent to the case of two resistors placed in parallel. If the spacing between the electrodes is doubled, again making a square, the resistance doubles, as for two resistors in series. The total resistance thus returns to the value it had when measured with the smaller square.

ous metal film breaks up into small islands as the plastic film substrate expands when heated. The effective conductivity of the sheet decreases, limiting the heating and thus the temperature. Second, the susceptor film is designed with a resistivity at a point below the maximum absorbing or heating point, say 75 Ω/sq. When the susceptor heats, its resistivity increases toward its most effective absorption range, around 188 Ω/square. Temperature then self-limits as its resistivity increases further and absorption decreases. This illustration can be followed by referring to Figure 7-1. Another potential problem, leaching of chemical toxins from the susceptor into the food due to the high temperatures reached, is covered in Chapter 9.

Most susceptors are fabricated by vacuum metallization where a thin film of aluminum is deposited on a PET film substrate. In this process, rolls of PET film are placed in a large vacuum chamber. As film is wound from a supply roll onto a takeup roll, it passes over a heated source of molten aluminum. The aluminum vapor, uninhibited by air molecules, is deposited on one side of the film forming the susceptor.

Recently susceptors have been fabricated from other materials. Nichrome and titanium nitride have been used with success. These specialty films are usually formed by *sputtering*. This process is also carried out in a vacuum, but a small amount of gas, such as argon, oxygen, or nitrogen, is introduced into the chamber in the vicinity of the film. The gas is ionized by high-voltage electrodes (microwave or radio-frequency energy can also be used) to form a plasma. The phenomenon is similar to that which produces an arc or lightning bolt (cf. Chapter 1), but the gas density is controlled so that no arcing occurs, only a glow similar to that found in a neon bulb. The plasma, which contains highly accelerated molecules, is directed onto a sample of the material to be used for the film coating. The high-energy ions or molecules bombard the sample and knock or sputter its atoms from its surface. These sputtered atoms stick to the film to produce the susceptor. Sputtered susceptors are usually more costly than vacuum-deposited ones and are not as widely used (Walters 1990).

Techniques have been developed to provide variable or controlled heating susceptors by selectively depositing conductive films in various patterns onto films (Walters 1991a,b) or by selectively etching away portions of a continuously deposited film (Wilson 1990). Providing a varying heating pattern of a susceptor is an excellent way to compensate for nonuniform microwave heating of a product.

Considerable effort is being made to fabricate printed susceptors. Potential advantages of this technology are cost and heating pattern flexibility. Incorporation of the susceptor into the package graphics offers potential unique competitive opportunities.

Flexible susceptors have been developed that allow food products to be wrapped. The intimate contact provided allows browning and crisping of all surfaces. Rolls of flexible susceptor materials, packaged for the consumer market for in-home use, have been tried successfully in many countries. To date this product has not met with customer acceptance in the United States.

Packaging configurations utilizing susceptor heater boards have been limited only by the designer's imagination. Susceptor boards covering the top and bottom of products have been successful, as well as susceptor tubes and sleeves. A unique package has been developed by Golden Valley Microwave Foods for microwave french fries. In this package individual fries are inserted into separate compartments of a square-cross-section "honeycomb" susceptor structure, crisping each potato on all four sides. A concern for most food product developers is the cost-to-value ratio when using susceptors in a package. Susceptor configurations are shown in Figure 7-2. Newer susceptor technology has been reviewed by Walters (1991b).

MICROWAVE ACCESSORIES

The first microwave accessory was the browning dish or "microbrowner" described previously. Tin oxide, a conductive film, was deposited on the back side of a glass ceramic dish. The conductive film interacts with the microwave electric field to produce heat. These microbrowners lost favor to less expensive metal-backed dishes, because of the high cost of glass ceramic. These less expensive versions were known as browning skillets or pizza crispers and used magnetic materials such as ferrite to interact with the microwave magnetic field. Magnetic susceptor material is particularly suited for use on a metal substrate since the electric field parallel to a metal boundary goes to zero while the magnetic field becomes a maximum. Conductive susceptors such as tin oxide and aluminum film must be used on an insulative substrate, since on metal the electric field is zero and there is minimal interaction or heating.

Popcorn poppers became the next popular accessory. Bulk popcorn placed in the popper quickly and conveniently provides a satisfactory product. Poppers quickly disappeared, however, with the advent of popcorn packaged in its own popping bag. During the early 1980s a wide variety of accessories flourished, with an estimated $75 worth being purchased with each microwave oven! Coffee makers, pressure cookers, waffle makers, griddles, bacon racks, and even microwave woks have been introduced to the market with varying degrees of success. Accessories are also illustrated in Figure 7-2. The popularity of accessories has continued to decline, primarily due to the incorporation of accessory functions into the package.

FIGURE 7–2.   Microwave Accessories and Packages: from upper left, clockwise: pressure cooker, popcorn popper, coffee maker, waffle iron, pizza on susceptor board, snack in susceptor sleeve   (Courtesy of The Rubbright Group).

References

Andreasen, M. 1988. New technologies to improve susceptor efficiencies in microwave packages. Proceedings of Pack Alimentaire '88, San Francisco, CA.

Bakker, M. (ed.). 1986. *Wiley Encyclopedia of Packaging.* New York: Wiley.

Blew, A. 1989. A resistivity measurement instrument to predict susceptor behavior. Paper read at MW Food Conference, 1989, Chicago, IL. (Available from Lehighton Electronics, Inc., Lehighton, PA.)

Brastad, W. 1980. *Method and Material for Preparing Food to Achieve Microwave Browning.* U.S. Patent 4,230,924.

Buffler, C. 1991. A simple approach to the calculation of microwave absorption, transmission and reflection of a microwave susceptor film. *Microwave World* **12**(3):5–7.

Cahners 1989. *Packaging Encyclopedia and Technical Directory.* Des Plaines, IL: Cahners Publishing.

Cahners 1991. *Packaging Sources Guide.* Des Plaines, IL: Cahners Publishing.

Chen, C. 1973. Transmission of microwaves through perforated flat plates of finite thickness. *IEEE Transactions on Microwave Theory and Technique* **Jan:**Vol.MTT-21:1-6.

Hanlon, J. 1984. *Handbook of Packaging Engineering,* 2nd ed. New York: Mc-Graw-Hill.

Harrington, R. 1987 (reissued). *Time Harmonic Electromagnetic Fields.* New York: McGraw-Hill.

Huang, H. 1990. Specifying and measuring microwave food packaging materials. *Tappi Journal* 73(3):215-218.

Keefer, R. 1987. *Microwave Heating Package and Method.* U.S. Patent 4,656,325.

Lampi, R. 1977. Flexible packaging for thermoprocessed foods. *Advances in Food Research* **23:**305-428. New York: Academic Press.

Lindstrom T. 1990. Evaluating the comparative performance of microwave susceptors. Proceedings of Pack Alimentaire '90, San Francisco, CA.

Marcuvitz, N. 1986. (reissued on behalf of Institute of Electrical Engineers). *Waveguide Handbook.* London: Peter Peregrinus. (Available through IEEE Service Center, Piscataway, NJ.)

McCormick, J. 1991. Historical and recent attempts to solve the uneven microwave heating problem with packaging. Paper read at TAPPI, San Diego, CA. (Available through AD Technology, Taunton, MA.)

Moffett, F. 1952. *Method of Heating Frozen Food Packages.* U.S. Patent 2,600,566.

Otoshi, T. 1972. A study of microwave leakage through perforated flat plates. *IEEE Transactions on Microwave Theory and Technique* **MTT-20:**235-236.

Packaging. Des Plaines, IL: Cahners Publishing.

Pesheck, S. 1990. The interaction of microwaves with susceptor and product. Microwave Packaging Symposium, American Management Association, Philadelphia, PA.

Ramey, R. and Lewis T. 1968. Properties of thin metal films at microwave frequencies. *Journal of Applied Physics* **39**(3):1747-1752.

Ramey, R., Kitchen, W., Jr., Loyd, J., and Landes, H. 1968. Microwave transmission through thin metal and semiconducting films. *Journal of Applied Physics* **39**(8):3883-3884.

Sacharow, S. and Brody, A. 1986. *Packaging: An Introduction.* Duluth, MN: Harcourt, Brace, Jovanovich.

Seiferth, O. 1987. *Food Receptacle for Microwave Cooking.* U.S. Patent 4,641,005.

Turpin, C. 1989. Browning and crisping: The functions, design and operation of susceptors. *Microwave World* **10**(6):8-13.

Turpin, C. and Perry, M. 1990. *Microwave Heater and Method of Manufacturer.* U.S. Patent 4,904,836.

Walters, G. 1990. Microwave susceptors: aluminum versus alloy coatings, optical density versus direct resistance measurements. Paper read at Society of Vacuum Coaters, St. Louis, MO. (Available from A. D. Technology, Taunton, MA.)

Walters, G. 1991*a*. Deficiencies of standard microwave susceptors in contrast to patent pending membrane susceptors. Paper read at Society of Vacuum Coaters, St. Louis, MO. (Available from A. D. Technology, Taunton, MA.)

Walters, G. 1991*b*. History and description of deficiencies in standard microwave susceptor films compared to 1990 developments in new susceptor performances. Paper read at Pack Alimentaire, New Orleans, LA. (Available from A. D. Technology, Taunton, MA.)

Welch, A. 1955. *Microwave Heating Apparatus and Method of Heating a Food Package.* U.S. Patent 2,710,070.

Wilson, D. 1990. *Demetallization of Metal Films.* U.S. Patent 4,959,120.

8

Microwave Product Development

The establishment of a microwave development facility requires several technical disciplines. The primary discipline is usually that of the researcher, whether food science or some other area. This chapter is designed to help the researcher establish a viable microwave laboratory and to describe the various considerations to produce a marketable microwave product. The descriptions are oriented toward the food product developer, but most of the techniques are applicable to any scientific or development program where microwave technology is considered.

The microwave laboratory for food product development is no different from a normal food product development laboratory except for the addition of microwave-specific equipment. Instrumentation required for a well-equipped microwave laboratory is summarized later. Fundamentals and technical details have been presented previously.

CHARACTERIZING THE MICROWAVE OVEN

The question most frequently asked is, how many and what type of microwave ovens are required for developing and testing a product? The answer is difficult because the population of ovens in use by the consumer is ever-changing. Early ovens were mostly high-power (greater than 600 W as measured by the 2-L test). In the early 1980s came an influx of inexpensive, low-power ovens. Recently, the tide has turned somewhat, with an increase in the higher-power ovens being seen. A few models are appearing with 2-L powers of over 800 W. A guess is that the present oven population consists of 60% high-power and 40% low-power ovens. Any facility must have representative models of ovens from all configurations sold on the

98

market (Schiffmann 1987, 1991). Table 8-1 proposes that at least six and preferably nine ovens are needed for a well-equipped laboratory.

To properly develop a product, the characteristics of each oven in the laboratory must be known and understood. Following are the most important oven characteristics with a description of how they affect product development. Notation in parentheses indicates the importance of performing the test. It is strongly recommended all information characterizing an oven be kept in a large envelope affixed to the side of the oven. This procedure eliminates lost data that always occurs when personnel and laboratories change.

Line Voltage (critical)

Line voltage determines proper operation of the microwave oven and ensures repeatability of power performance. To ensure that the power output of laboratory ovens remains consistent from operation to operation, it is

Table 8-1   Minimum and Recommended Ovens for a Food
Product Development Laboratory*

Less than 500 W[†]

Minimum: none
Recommended: 1, for comparison purposes only.

Low Power (500-600 W)

These ovens usually have cavity volumes ranging from 0.5 ft³ (0.014 m³) to 0.8 ft³ (0.023 m³).
Minimum: 2; 1 with stirrer, 1 with turntable if available, spanning cavity size range.
Recommended: 3; with stirrer and turntable and spanning cavity size range.

High Power (600-800 W)

Cavity volumes are usually between 0.9 ft³ (0.026 m³) to 1.3 ft (0.037 m³).
Minimum: 3; at least one with stirrer and one with turntable and spanning cavity size range.
Recommended: 4; 2 with stirrer, 2 with turntable spanning cavity size range.

Combination Microwave/Convection

Minimum: 1
Recommended: At least 1, with metal turntable. It may be appropriate to specify that products not be used on metal turntables. A fully equipped laboratory may have a second optional combination oven with glass or ceramic turntable.
Total Minimum: 6
    Recommended: 9

*Power notated is measured by 2-L test. Add 100 W to convert to IEC power.
[†]It may be appropriate to specify on label that products are not designed to be used in ovens with power less than 500 W.

mandatory to know if the line supply voltage remains constant. Line voltages from laboratory to laboratory, even in the same building, can vary from 117 to 127 V. Similar variations can occur in a single laboratory during a day, especially if the laboratory is located near a manufacturing facility where heavy electrical machinery is used. Line voltage changes can cause unacceptable variation in oven power (Figure 2-7).

As a minimum, voltage should be monitored hourly during the working day for several days during the week before operating the ovens in the laboratory. A company electrician can provide a voltmeter that can be monitored hourly or, preferably, a recording voltmeter that will provide a continuous record throughout the day; or a multimeter can be bought at an electronics store. A plug should be wired to the multimeter test leads by a person experienced with electrical equipment. The meter should be plugged into all receptacles (wall sockets) that will be used for the laboratory microwave ovens, and data recorded.

CAUTION. The tests leads of some inexpensive multimeters have their ends that plug into the meter exposed. This situation is hazardous and could result in injury from electrical shock if the leads are accidently left plugged into the wall socket and removed from the meter. *Use only multimeters with shielded test lead ends,* such as Fluke model 12.

A better procedure is to use a monitoring box that measures line voltage as well as oven current. The box has a variable transformer that allows adjustment of the line voltage to the proper value if the voltage changes. Units are available commercially (Appendix 4). A circuit diagram is shown in Figure 8-1. Monitoring oven current is useful for determining magnetron turn-on time (Figure 2-8*b*) or calculating oven efficiency. Oven efficiency,

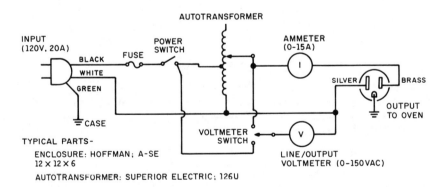

FIGURE 8-1.   Power Monitor Circuit Diagram.

e_{ff}, is the ratio of oven output power, P_o, (Equation 4–4) and oven input power, P_i, given by the line voltage V multiplied by line current I:

$$e_{ff} = \frac{P_o}{P_i} = \frac{P_o}{VI}$$

$$(8\text{–}1)$$

Oven Power (critical)

Oven power determines how rapidly a product heats in the microwave oven (Chapter 4). Two-liter and IEC (Appendix 5) powers should be measured to determine the correlation between the two. IEC power is typically 8–12% higher than 2-L power. On rare occasions it may be less. A one-cup (U.S.) (237-mL) time-to-boil test should also be performed if recipes or directions are being developed using "high" and "low" categories (Chapter 4). The three power tests are described in Appendix 5.

Power versus Initial Operating Time (important)

Power versus operating time determines the power dropoff of a cold oven when it is initially operated (Chapter 4). A 2-L or IEC test may be used for this measurement. For simplicity, however, a 1-L test can be used as long as it has been previously correlated to either the 2-L or IEC test. The test is most efficiently done with two persons and two beakers. Person 1 fills a beaker and measures initial water temperature. Person 2 places the beaker in the oven and runs it for the prescribed time while person 1 readies the next beaker. Person 1 measures and records the final water temperature while person 2 cools the shelf between runs by wiping it with a wet rag and resets the timer for the second run. In this manner, no delay between measurements is experienced. Four or five sequential measurements should be made and plotted.

Power versus Load Volume (important)

Power versus load volume determines power coupling efficiency to small loads (Chapter 2). The 2-L power measurement technique (Appendix 5) with Equation 4–3 is most conveniently used to make this measurement. Two 1-L beakers, one 2-L beaker, or the IEC crystallizing dish (Corning) may be used. Measure the temperature rise in 2000 mL of water for 2 min, 1000 mL for 1 min, and 500 and 250 mL for 30 s and plot results. Calculate the ratio r, of the 500-mL power to the 2000-mL power. For an oven with adequate coupling efficiency, r should be greater than 0.80 (Buffler 1990). Note that the use of English units is not specified nor recommended for these tests.

Oven Pattern (optional)

The oven pattern determines if the oven has extraordinary hot spots. A gross idea of the location of hot and cold spots within an oven can be obtained with a simple one-dimensional mapping technique. By placing a layer or sheet of thermally sensitive material inside an oven, one can produce a map of hot spots. Numerous materials have been used, such as wet sand, sawdust, cornmeal, pie dough, liquid crystal sheets, or computer paper. Thermal paper is by far the easiest to use. Infrared paper (3M), or any thermal computer printer paper, is placed inside the oven. The paper must be inserted inside a plastic sheet protector (found in any stationery store) along with the paper insert that comes with the protector. The thermal paper by itself will not heat; it must be in contact with the paper insert to produce a pattern. For sheet protectors stored in an air-conditioned environment an exposure time of 30 s should be sufficient. With humid storage, as little as 5 s may be sufficient. Experimentation might be necessary to avoiding scorching of the paper.

Sheets placed at various heights indicate three-dimensional hot spots. Excellent spacers for microwave oven use are white expanded polystyrene cups. These cups are virtually invisible to microwaves and are excellent heat insulators. The drawback is that they melt at slightly over 100°C (212°F).

There is very little correlation between the pattern shown on a one-dimensional sheet and overall oven cooking performance. Any food load radically alters the heating pattern from that found in an empty oven. The sheets are valuable for determining if there are very intense hot spots.

Time Base and Duty Cycle

Time base and duty cycle determine pulse duration for variable power settings. (These parameters are important if product instructions are being developed that require a lower-than-full power setting.) If a monitoring box is available (Figure 8–1), the time base and duty cycle can be calculated by measuring the on-time—the time that the current meter registers full current—and the off-time—the time at which the current is minimal. The time base is the sum of these two times. The duty cycle is the ratio of on-time to time base (cf. Figure 2–9 and Equations 2–1 and 2–2).

Without a monitoring box, the parameters can be measured by means of a standard 15-in- (5.9-cm) long, $1\frac{1}{4}$-in- (5-cm) diameter (model series F14T12) fluorescent lamp. The tube should be placed in the oven adjacent to or on top of a measuring cup filled with approximately 1 L of water. Exercise caution to keep the pins on the end of the bulb away from cavity walls because they might arc. The on- and off- times of the flashing light are measured and used in Equations 2–1 and 2–2.

CAUTION. The bulb rapidly becomes very hot! Turn the oven on only long enough to make the measurement. Leave the bulb in its original paper sheaf if possible; otherwise wrap it with cardboard. Remove from oven cautiously!

As a last resort, if the voltage-measuring meter is available, it may be possible to monitor the on-off times by the drop in voltage when the meter is plugged into the receptacle on the same wall plate as the oven. Typically a drop of about 3 V occurs when the magnetron is on. With some ovens it may be possible to time the on-off "humming" sound of the transformer or the dimming of the oven light.

MEASUREMENT OF PRODUCT PARAMETERS

In many circumstances it is desired to monitor product parameters while it is undergoing microwave heating. Due to the unique properties of microwaves, the usual techniques may not be applicable. Temperature, pressure, and electric field measurements are of such a nature. The technologies for such measurements were discussed in Chapter 3, as applied to sensors in a microwave oven. A brief discussion of these technologies as applied to experimental parameter monitoring of heated products is given next. A list of instrumentation manufacturers is given in Appendix 4.

Temperature

Conventional techniques, such as thermocouples, for monitoring temperature in a microwave oven are difficult, if not impossible, to implement reliably. The need to use wire leads invariably produces arcing and burning. Wire leads also perturb the electric field and distort the heating pattern. Also, modern electronic thermocouple monitors can be severely disrupted or damaged by microwave interference transmitted by the leads into the instrument.

Temperature measurement under microwave conditions has been well solved by the use of fiber-optic sensors. The fibers are small and virtually invisible to microwaves. They also produce negligible thermal distortion in the product measured. Different technologies have been utilized, but they all use a material on the end tip of the fiber that has an optical property that changes with temperature. The tip material is interrogated by a pulse or beam of light propagated down the fiber from within the instrument. The light returned from the temperature-sensitive tip is compared with the incident light and is analyzed to extract temperature information. Instrumentation with up to 12 channels has been developed. An overview of fiber temperature-sensing technology has been prepared by Wickersheim and Sun (1987).

To allow entry of the fibers into an operating oven, it is safe to drill or punch small holes in the oven cavity and external wrap. Holes of $\frac{3}{8}$ in (0.95 cm) diameter are sufficient to allow entry of multiple fibers. Since the fibers are fragile, it is advisable to line the holes with a short length of plastic tubing. Be careful not to allow any metal to be inserted into the hole because it will act as an antenna and cause arcing, leakage, and possible burns. Leakage may be checked with a survey meter (Chapter 9) but should be well under 5 mW/cm^2.

To protect and support the fibers inside the oven as they are inserted into the product, use a plastic holder. Fluorocarbon resin, such as Teflon (registered trademark of Du Pont), is the preferred material for the stand because it has a loss factor less than 0.001 at 2450 MHz. Holes in the holder should be slightly larger than the fiber to allow easy passage. The fiber may be secured with a small piece of tape if needed.

Pressure

For applications where pressure needs to be determined, such as superheating and high-temperature processing (sterilization), fiber-optic technology has been developed. In such instrumentation, a miniature sealed cavity is affixed to the end of an optical fiber. The terminal end of the cavity consists of a diaphragm exposed to the environment. A change in external pressure with respect to the pressure inside the cavity causes a deflection of the diaphragm. The cavity diaphragm is interrogated by a light beam transmitted through the fiber from the instrument. The change in the spectral quality of the reflected signal with respect to the transmitted signal is a measure of the diaphragm deflection and, thus, pressure. A survey of the technology has been presented by Saaski, Harti, and Mitchell (1986).

Thermal Imaging

For the development of many products or processes it is desirable to determine the temperature distribution in a product while it is heating. Thermal imaging is possible through the use of infrared (IR) camera systems. These cameras are similar to videocassette recorders except that they require cooling to make them sensitive at infrared wavelengths. Cooling is often implemented by compressed gas contained in external tanks. The thermal images produced are displayed on a color monitor, with different colors indicating the temperature ranges of the product surface. Software packages are available that analyze the image and provide statistical data, such as area-versus-temperature histograms, average temperature, range, and standard deviation.

For these cameras to view the product during heating, the microwave oven to be used must be modified. Usually, adequate images can be obtained by viewing through the screen in the oven door if the plastic sheet or film covering the screen is removed. Plastic transmits infrared poorly and thus distorts a thermal image. Since all doors are different, each must be examined to determine plastic removal technique. In many instances the oven wrap must be removed to loosen the door hinge to obtain access to the plastic sheet. The oven should be modified only by an experienced technician. A superior image is obtained by removing the metal screen and replacing it with a copper window screen, secured by solder, if possible. Copper or silver conductive cement can be used and then covered with aluminum tape. The disadvantage of copper screen is its potential to arc where attached to the door, especially if the oven is operated with a small load.

CAUTION. Following door modification a leakage measurement *must* be made (Chapter 9) with an approved survey meter (Appendix 4). Follow-up measurements should be made after each week of oven operation.

Since the screen acts as a filter to infrared, calibration of the oven is required. A hot plate painted black or with a black block of copper placed on it is required. A conventional thermocouple is securely attached to measure actual temperature. The hot plate is then viewed through the screen by the IR camera with the door open. A calibration curve is then made of the thermographic temperature measured by the camera versus the actual temperature measured by the thermocouple as the hot plate temperature is increased. **Caution!** Do not place the hot plate inside the microwave oven nor operate the oven while making this calibration.

Most materials are neither perfect absorbers nor reflectors of IR radiation. The degree of imperfection is measured by emissivity from 0 to 1 (0% to 100%). Perfect absorbers, called IR blackbodies, absorb all thermal energy incident on them and reflect none. They have an emissivity of 1. Perfect reflectors of thermal energy absorb none and have an emissivity of 0. Most foods have an emissivity of approximately 0.85. For most measurements this value suffices as an additional correction to the screen factor. For more accurate measurements, sample emissivity as well as the screen correction should be accounted for by heating the actual product in the oven, viewing it with the IR camera, opening the oven door and then probing the surface of the sample with a thermocouple probe to determine temperature. **Caution!** Do not operate the microwave oven with the thermocouple in the sample. Probe the sample only after the door is opened.

Often it is more economical to use a substitute or mimetic of a product rather than the product itself. In this fashion, viewing the effects of geometry or dielectric changes may be readily accomplished. Agar, for example, can be mixed with salt, to change its dielectric properties, and molded into any desired shape. Instant mashed potatoes may even be used. Another mimetic is TX 151 from Oil Center Research (cf. Appendix 4), with dielectric properties similar to many foods. Research on the absorption of microwaves by human body parts has led to extensive research on recipes for mimetic materials (Chou et al. 1984).

Since IR cameras are extremely expensive, a compromise technique for obtaining an idea of temperature patterns within a product can be used. The product is first heated in a microwave oven for a given time. The product is then removed and probed with an array of conventional needle-type thermocouples. Any number of thermocouples can be arrayed by securing them in holes drilled in a thin metal plate. The array is then attached to an inexpensive device (found at any hardware store) that turns an electric drill into a drill press. The drill press with thermocouples can be quickly inserted into the product by simply pulling down the handle. Presetting the probes' positions and depth, as well as the distance the drill press descends, allows accurate and repeatable measurements. Data-logging equipment is available as stand-alone instrumentation or as hardware and software, which is used in conjunction with a personal computer (Appendix 4).

PRODUCT TESTING

Developing, testing, and evaluating microwavable products demands at the very least the same care and protocol bestowed on conventional products. In addition, due to the nature of the microwave oven, certain additional procedures should be observed.

1. All ovens used for testing should be characterized as described above.
2. In addition, further recommendations and details have been presented (Schiffmann 1987, 1990; Stanford 1990):
   a. Before ovens are used they should be warmed for 10 min with a large water load. This procedure stabilizes power output and is especially important for ovens with large power dropoff (cf. Section 3 above).
   b. Place samples at the same location from test to test.

   c. Repeat handling instructions exactly from test to test.

   d. Rotate tests between ovens to prevent overheating of a single oven.

   e. Check oven power before each day's use. If power varies by greater than 5%, check line voltage and retest.

Instruction Tolerance Determination

The most critical procedure in instruction and recipe development is tolerance testing, because of the short preparation time involved and the wide variability of available oven powers and coupling factors. The power of ovens on the market ranges from below 300 W to 800 W (2-L test) or from 400 W to 900 W (IEC powers). Many microwave instructions do not list preparation time as a function of wattage, but break power into two categories, say "high," greater than 600 W, and "low," less than 600 W (2-L test). In the high range power can vary up to 33% from the upper to lower limit. Low-power ovens can vary by 100%! Since power is related to cooking speed, time to reach optimal product temperature can vary by this much. Because of the possible large variation in cooking speed of low-power ovens and their poor heating characteristics, it is advisable to consider excluding ovens under 500 W on instructions. With these ovens eliminated, preparation times vary by only 20% for low-power ovens.

It is extremely important to determine the organoleptic qualities as a function of preparation time. If the product is acceptable over a ±15% (30% total) variation from optimal preparation time, when tested in a high-power oven, it will probably provide acceptable quality when prepared in most high-power ovens. Similarly, if a 20% time range provides acceptable results when tested in a low-power oven the product should also work in most low-power ovens. Instructions for products having such wide tolerances should be optimized for preparation time in an oven with a power approximately midway in each range (i.e., 550 W and 700 W). If the product is more sensitive to time or power variation, it should be optimized at powers near the bottom of the range. In this way the consumer will undercook on the first try but will continue to add small increments of time until the product is done. It is certainly advisable to attempt to reformulate the product, change its geometry, or introduce unique packaging in order to increase the tolerance to acceptable levels if at all possible.

The dividing line between high and low power is arbitrary and may be chosen by the manufacturer. With the advent of dual-power tests it is advisable to indicate 2-L or IEC on the instruction. In the future, a changeover to IEC will probably occur. In the meantime the 2-L test covers 98% of the ovens in use.

Abuse Testing

The most forgotten consideration in developing a microwavable product is determining what would occur to the product in a worst-case scenario. On extremely rare occasions a product might be left accidently in an oven with the power left on. Products could catch fire, burn, erupt, scald, or produce a variety of barely imaginable problems. As a final test procedure, it is suggested that a "blue sky" or ideation session be held to uncover any possible mishaps. Chapter 9 should be reviewed for various safety hazards. Testing should be performed on all ideas generated to determine actual results. A risk assessment review with management is then held before the product is launched.

References

Buffler, C. 1990. An analysis of power data for the establishment of a microwave oven standard. *Microwave World* 11(3):10–15.

Chou, C., Chen, G., Guy, A., and Kuk, K. 1984. Formulas for preparing phantom muscle tissue at various radio frequencies. *Journal of Bioelectromagnetics* 51:435–441.

Corning. Crystallizing Dish Model 3140–190. Corning Glass Works, Corning, NY.

3M. Infrared Copy Paper, Type 0111. 3M Company, Office Systems Division, St. Paul, MN.

IMPI 1991. International Microwave Power Institute. Report of standards committee.

Saaski, E., Harti, J., and Mitchell, G. 1986. A fiber optic system based on spectral modulation. *Advances in Instrumentation* 41(3):1177–1184.

Schiffmann, R. 1987. Performance testing of products in microwave ovens. *Microwave World* 8(1):7–13,15.

Schiffmann, R. 1990. Microwave foods: Basic design considerations. *Tappi Journal* 73(3):209–212.

Schiffmann, R. 1991. Oven considerations for the food product developer. Paper read at the 26th Annual Symposium, International Microwave Power Institute, Buffalo, NY.

Stanford, M. 1990. Microwave oven characterization and implications for food safety in product development. *Microwave World* 11(3):7–9.

Wickersheim, K. and Sun, M. 1987. Fiberoptic thermometry and its applications. *Journal of Microwave Power* 22(2):85–94.

9

Microwave Safety
and Regulations

The microwave oven has many times been described as the safest appliance in the kitchen. Recently it has even been mentioned as the method of choice for food preparation because it does not produce the carcinogens associated with browning and charring. Providing a safe environment for microwave cooking, as with all food preparation, requires supervision where necessary and common sense. Microwave technology, being relatively new, does, however, require a few considerations unique to this type of cooking.

Microwave safety issues can be broken down into four categories: interference to communication and electronic systems; exposure of oven users to microwave radiation; the safety of the cooking or preparation process; and the safety of the prepared or processed food. These topics will be dealt with individually.

ELECTRICAL INTERFERENCE

Federal Interference Regulations

The generation of microwave power for use in microwave cooking and processing is well regulated in the United States by the Federal Communications Commission (FCC), an independent organization reporting directly to the U.S. Congress. Requirements are documented in the *Code of Federal Regulations* (FCC 1991*a*). The frequencies now allowed for microwave cooking and processing were originally allocated in 1946 for use by the industrial, scientific, and medical (ISM) communities. With the advent of practical microwave cooking appliances in 1958, the ISM frequencies were allowed for such use (Osepchuk 1984). ISM frequencies and their uses are listed in Table 10–2. Heating at frequencies below 915 MHz is generally referred to as dielectric or radio-frequency (RF) heating. A review of the types of electromagnetic heating has been published by Metaxas (1988).

Aside from the allocation of the primary or fundamental frequencies for use by the microwave community, the FCC also limits levels of emissions at other frequencies that might cause interference with other services, such as communications, or other equipment. Manufacturers of microwave ovens sold in the United States must be able to verify that the spurious emissions from their ovens fall below FCC limits. This verification is performed at the oven manufacturer's site or by a qualified independent testing laboratory.*

Verification of microwave processing equipment is more difficult than for microwave ovens. For processing equipment, the responsibility for verifying FCC acceptable emissions rests upon the site of installation. Alternatively verification depends on the manufacturer's assurance that his equipment meets FCC standards and that the emissions are independent of the installation. If this verification cannot be ensured, on-site testing must be performed.

Further information on electromagnetic interference and testing can be obtained through trade journals in the field (EMC 1991).

Pacemakers

Aside from the important practical problems of microwave ovens and microwave equipment potentially interfering with communications, microwave airport landing systems, and other equipment, the perennial question of whether microwave ovens interfere with pacemakers still exists. In the early days of pacemakers (e.g., pre-1968), many types of equipment and appliances could potentially interfere with pacemakers. Vacuum cleaners, mixers, engine ignition systems, and microwave ovens did emit spurious signals that could interfere with proper operation of pacemakers as well as television and radio sets. Since that time, two events have occurred that have essentially eliminated any danger to pacemaker users. First, the establishment of leakage standards in 1968 set maximum and known levels of microwave leakage from ovens. Second, with this information, pacemakers were redesigned with electrical shielding and filtering that prevented any low-level spurious signals from interfering with correct operation.

Today, while it is impossible to say that under every conceivable circum-

---

*Even though many microwave ovens are controlled by computer microchips, they are presently exempt from FCC regulations covering microprocessor-controlled electronics (FCC 1991*b*).

stance interference is impossible, pacemaker wearers need not worry. In fact, "warning signs can cause more undue apprehension in pacemaker patients than the potential risk merits" (Medtronics). A comprehensive review of the pacemaker issue has been published by Osepchuk (1981).

MICROWAVE LEAKAGE AND EXPOSURE

The association of low photon energy, nonionizing microwave radiation and high photon energy, and ionizing nuclear and x-radiation in the public mind is unfortunate. It is well documented that high-energy electromagnetic radiation produces damage to genes, cells, and tissue. The damaging, ionizing effect begins in the region of ultraviolet light, around 1.5 PHz. It is well known, for example, that exposure to sunlight causes skin cancer.

With regard to microwaves, to date there have been no authentically documented or verified incidents of health effects. To the contrary, beneficial results have been demonstrated by researchers who have used microwaves to treat muscle damage, by diathermy (Guy and Chou 1983, Ghandi 1990) and cancerous tumors, by hyperthermia (Cheung and Al-Atrash 1987). Also, the efficiency of using microwaves for incubation of mammals (Braithwaite et al. 1989), chickens (Braithwaite et al. 1991), and whole-body heating of people (Buffler 1988) has been shown with no ill effects.

Federal Leakage Regulations

It is well understood that body heating can occur if microwave power densities are sufficiently high, since this heating is the mechanism of microwave cooking and processing. To this point, in 1968 federal regulations were adopted that limited the emissions from microwave ovens and processing equipment (FDA 1991) to 1 mW/cm^2 measured 5 cm from any portion of the oven when removed from the packing box or 5 mW/cm^2 over the life of the oven. With the advancement of microwave oven technology, ovens now normally have leakage levels well below the required federal mandates. To put leakage levels in perspective, a comparison of the heating effects of various power densities of microwave and solar radiation is given in Table 9-1. These levels are also compared with microwave oven standards and performance.

Oven Leakage versus Exposure

It is often quoted in the press and assumed by microwave oven owners that leakage regulations in Russia and the old Eastern bloc countries are far more stringent than those in the United States. These beliefs engender fears

Table 9-1   Comparison of Microwave and Solar Radiation Levels

Radiation Level (mW/cm^2)	Remarks
300	Produces severe burns
110	Summer midday at equator*
70	Sunny day in summer, mid-United States*
50	2 ft from fireplace
20	Threshold of sensation on human body
5	Maximum FDA leakage allowance over life of microwave oven
1	Maximum FDA leakage allowance at time of sale
0.1	Typical oven leakage at manufacture

*Produces sunburn.

that more knowledge of adverse effects of microwaves may be available in these countries than in the United States. The reason for this misconception is that the U.S. regulations are for microwave leakage, while the former Soviet Union standards are for exposure.

The U.S. microwave leakage standard, as discussed, is a measure of the power density or flux emanating from an oven when measured 5 cm from any part of the oven. The strength of the leakage thus depends solely on the oven design. Exposure, on the other hand, is exactly that—the amount of radiation actually falling on the body over a specified time. Since microwave radiation decreases as the inverse square of the distance from the oven, the amount of flux to which the body is exposed is miniscule at any typical distance from an operating oven.

As an example of this square-law property, the leakage from an oven as measured at 5 cm (approximately 2 in) from an oven would produce an exposure reduced by a factor of $(2 \text{ in}/12 \text{ in})^2 = (5 \text{ cm}/30 \text{ cm})^2 = 0.03$, 1 ft (30 cm) from the oven, or a factor of 0.007 at 2 ft from the oven (60 cm). Thus, an oven with leakage within federal standards of 1.0 mW/cm^2 would expose a person to 0.03 mW/cm^2 at a distance of 1 ft (30 cm) from the oven. The exposure at a distance of 2 ft (60 cm) would be 0.007 mW/cm^2 (Figure 9-1).

The published former Soviet Union exposure level (Savin 1983) is to maintain a time-averaged exposure to the whole body of less than 0.01 mW/cm^2 for 24 h. If one integrates the microwave exposure to an oven user as a function of distance from the oven as well as the time-exposed, one immediately sees that exposure from an oven meeting U.S. leakage standards falls below the exposure standard required by the former Soviet Union.

In the United States there is as yet no federal microwave exposure regulation. The Occupational Safety and Health Administration (OSHA) does,

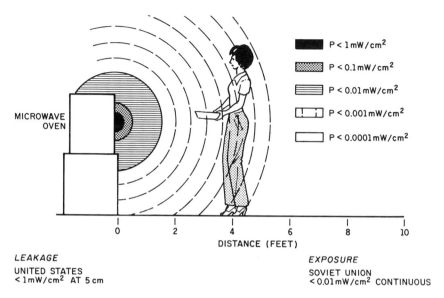

Legend:
- P < 1 mW/cm²
- P < 0.1 mW/cm²
- P < 0.01 mW/cm²
- P < 0.001 mW/cm²
- P < 0.0001 mW/cm²

MICROWAVE OVEN

DISTANCE (FEET)

LEAKAGE
UNITED STATES
< 1 mW/cm² AT 5 cm

EXPOSURE
SOVIET UNION
< 0.01 mW/cm² CONTINUOUS

FIGURE 9-1.   Microwave Leakage and Exposure   (Adapted from Osepchuk 1973).

however, provide a guideline for exposure in the workplace of less than 10 mW/cm² for any 6-min period (OSHA 1990). This guideline is currently under review. Most workplaces, in the interim, accept the whole-body standard established by the American National Standards Institute of 5 mW/cm² (ANSI 1982).

Leakage Monitoring

The extensive experience amassed over the last two decades of microwave oven manufacture, with an estimated 116 million units having been sold (Packaging Strategies 1989), indicates that the possibility of having an oven leak over the amount allowed by the standard is extremely remote. Care should always be exercised, however, to see that an oven is not operated with any object caught in the door. If the door does not close properly, or if there is damage to the door, hinge, latch, or sealing surface, it should not be operated but and taken immediately to authorized service personnel for testing and repair.

    If there is no physical evidence to indicate that an oven may be leaking and there is still a concern, several options are available. First, since virtually all appliance centers that service microwave ovens have a leakage-monitoring survey meter, it would be appropriate to contact them to determine their policy for measuring oven leakage.

Inexpensive leakage survey meters are on the market in electronics stores in almost all cities and cost from $5 to $50. It is *not* recommended that these be purchased, since they have been determined to be extremely inaccurate. A study done by the Food and Drug Administration (HEW 1979) reports that these meters can indicate values that are either much higher or much lower than the actual leakage.

Accurate leakage survey meters which have been approved by the Food and Drug Administration's (FDA) Center for Devices and Radiological Health are available starting at $199. Manufacturers are listed in Appendix 4.

A number of states have departments of radiological health that will perform oven leakage measurements if they determine that there is a potential for a health hazard from a leakage exposure. A call to the state governor's office should locate the appropriate department.

In years past, local FDA offices of the Department of Health and Human Services would measure microwave oven leakage if they deemed the specific situation warranted an inspection. At present, due to limited staff and resources the FDA only investigates events that they feel represent a serious potential hazard, and only if a formal complaint is filed. The United States, Puerto Rico, and the Virgin Islands are divided into six regional FDA offices: Brooklyn, Philadelphia, Chicago, Atlanta, Dallas, or Seattle. These offices may be contacted for information.

HEATING AND COOKING SAFETY

Even though it has been said that the microwave oven is the safest appliance in the kitchen, precautions should be observed by the prudent user. Some of these follow common sense guidelines, while others may not be obvious until the unsuspecting user experiences their first accident. Discussion of microwave oven safety issues have been well detailed by Osepchuk (1978).

Superheating, Eruptions and Explosions

Differential heating by microwaves has been shown to cause hot and cold spots in a product. Usually this phenomenon is simply annoying, causing less than optimal product quality. Under certain circumstances, some liquid or flowing products may exhibit a phenomenon of eruption caused by superheating, especially if recipe or instruction times are inadvertently exceeded.

Susceptible items include stews and viscous soups, particularly if they are heated in a vessel with vertical walls, such as a mug, cup, or deep bowl.

With these products, internal hot spots may develop due to a combination of geometry, dielectric properties, (Chapter 6), and oven field nonuniformities (Chapter 2). As heating proceeds, internal steam is produced. For viscous media in a straight-walled container, convection flow is inhibited and steam pressure builds until it is suddenly released in an eruption. This eruption, sometimes called volcano effect or "burping," virtually blows the product out of the container. In most cases, only a soiled oven results. In some circumstances the eruption occurs when the product is being removed from the oven, which could burn the consumer.

Nonhomogeneous products, especially those with large particulates, sometimes exhibit the same phenomenon. The differential heating, and subsequent eruption, of these products occurs because of the preferential heating of the particulates due to their geometry and dielectric properties.

A much more rare phenomenon, but one that has been documented, is eruption caused by the superheating of low viscosity fluids such as water or thin soups. As discussed previously, hot spots may occur in the water being heated. If the temperature of the water reaches the boiling point, boiling usually occurs. But under rare circumstances, boiling could be inhibited and superheating occurs, with the temperature of the water at the hot spot rising as much as $18°F$ ($10°C$) above the boiling point. Under this condition the superheated water is unstable. Any motion, such as vibration or removal of the vessel from the oven or the addition of powdered coffee or tea, might cause sudden violent boiling with an accompanying eruption from the vessel. Superheating is promoted if there are no nucleating points such as dust or scratches on the vessel to initiate the boiling, hence the rapid boiling when coffee or tea is added. Dissolved gas coming out of solution in the form of bubbles when the water is heated could act as nucleating points. Thus, if there is little dissolved gas in the water, which could happen if warm or long-standing room-temperature water is heated, superheating might occur (Buffler and Lindstrom 1988).

Precautions to avoid burns while heating or boiling liquids in a microwave oven include

• Start with cold tap water if possible.
• Use a well-worn, not new, vessel.
• Use a cup with sloping sides in preference to a mug.
• Stir several times during heating.
• Add powdered coffee or tea to water before heating.
• Do not exceed the recommended heating time.

One of the most common mistakes made by the consumer is to try to prepare eggs in the microwave. The dielectric properties and the geometry of an egg contribute to an enhanced focusing effect. Since the egg yolk is contained in a membrane, any pressure buildup is exaggerated, causing the possibility of a severe explosion. Never hard-boil an egg directly in a microwave oven. Some cookbooks give instructions to hard-boil an egg by immersing in water while microwaving. Accessories are also available in which an egg is placed in a metallic shell and immersed in boiling water. Prudence, however, recommends that hard-boiling an egg be done on a stove top.

Reheating a hard-boiled egg in a microwave should not be attempted because the pressure buildup in the yolk sac may cause the egg to explode when cut or eaten. Frying an egg may be attempted if the yolk sac is pricked in several places before microwaving to prevent pressure buildup.

Burning and Scalding

Overheating any foodstuff could result in burning or scalding. With microwave heating, however, two circumstances require particular attention since overheating might not be obvious. Both of these concerns involve feeding babies.

Baby bottles heated in the microwave oven might exhibit internal hot spots. Since the heating time for the bottle is relatively short, the glass or plastic bottle may not have had time to heat appreciably and might feel cool. The care provider might not then be aware of the scalding temperature of the bottle contents when giving it to the baby. If the microwave oven is used for heating formulas, extreme care must be exercised. The bottle should be well shaken and tested on the back of the hand before use.

A similar phenomenon could occur when a jar of baby food is heated in the microwave oven. The jar should be well stirred and tested before feeding.

Other areas of precaution against burning and scalding include

- Steam venting from popcorn bags upon opening.
- Steam venting from plastic wrap or other covered dishes upon removal from oven.
- Hot objects such as tomato when placed on top of cheese and cracker heated in an oven.
- Hot insides of multicomponent foods such as jelly doughnuts and filled pastries.
- Use only microwave safe dishes; some dishes heat in a microwave oven and might cause burns.

Fires

Under normal cooking or heating procedures in a microwave oven, most of the product is limited to the boiling temperature of water as the water is evaporated. Fats or foods with very high fat content are exceptions and continue to heat until the smoke point or, possibly, the flash point is attained.

If under abusive circumstances a product is heated until the water is evaporated, two situations may exist. If the microwave absorptivity (dielectric loss factor) of the dry constituents that remain are low, microwave heating declines, with temperatures remaining below the boiling temperature. This circumstance is called the self-limiting effect. If the remaining dry material has a high microwave absorptivity, the product immediately starts to heat rapidly above the boiling temperature of water. Most food products contain carbohydrates that are good intrinsic absorbers of microwaves. Thus, desiccated foods heat rapidly to their burning point of 451°F (233°C) if left in the microwave oven. This type of heating desiccated products is called the *runaway* heating effect. The unique penetration-depth properties of some foods coupled with a spherical or cylindrical geometry can easily produce a focusing effect (Chapter 6). Foods such as potatoes, squash, or marshmallows, if left too long in the oven, could desiccate internally due to the internal concentration of microwave energy. Under these circumstances there is usually no external evidence of the internal runaway heating until the food smokes or bursts into flame.

The runaway effect can also be responsible for the burning of other articles placed in the microwave oven. It is not recommended, for example, to attempt to dry herbs, newspapers, shoes, or clothing in a microwave oven.

Critical precautions to take while microwaving all foods or food products are as follows:

- Set cooking time with extreme care. Digital timers are particularly susceptible to added zeros.
- In prepackaged microwavable foods, read the directions carefully. Always add the prescribed amount of water.
- Microwave cooking usually takes only a short time. Do not get distracted; stay close to the oven during microwaving.
- Always supervise children operating a microwave oven.

Metal in the Microwave

Most users of microwave ovens still believe that metal (e.g., aluminum foil or aluminum containers) should not be used in the oven. This concern might have been true 15 years ago; engineering improvements, however, have essentially eliminated the technical problems in modern ovens.

Previously, metal placed in the oven cavity reflected a portion of the microwave energy entering the cavity back into the magnetron. This energy was then dissipated as heat internal to the magnetron, causing distortion, and even melting, of the vanes and straps (Chapter 2). Excessive magnetron heating did cause oven failures or, at least, reduced oven life. Microwave ovens, today, however, utilize more ruggedly designed magnetrons that allow higher internal anode temperatures. The ovens themselves are also engineered so that less power is reflected back to the magnetron when there is no food load in the oven.

Food products in metal trays may be defrosted or heated with no associated problems. Improvement in product heating uniformity improvement has been reported (Decareau 1978), due to shielding edges and corners and from slower heating, which happens if microwaves enter from the top only. Microwave accessories for heating, browning, and grilling have been developed from metal. These products are sometimes accompanied by booklets designed to alleviate the consumer's concern (Raytheon 1982).

The use of foil to prevent overcooking sensitive parts of food products, for example, wings or legs of whole birds has been recommended for 20 years (Litton 1971). Use of large foil pieces, however, especially on large objects and near the top of the oven, is not recommended.

Arcing

The electric field in the microwave oven accelerates any charged molecules found in the cavity air. These accelerated molecules collide with their neighbors, and, if the field is high enough, sufficient energy will have been gained to knock their neighbor electrons loose. This process, called *ionization,* produces a positively charged molecule, or ion. The dislodged electrons could be carried off by the impacting molecule, or they could attach themselves to neutral molecules, thereby producing a negatively charged ion. If the procedure repeats itself over and over, an avalanche of flowing ions in the air occurs. This flowing current of ions is a "lightning bolt" if it occurs in nature or an arc if it occurs in a microwave oven.

Magnetron power cannot provide a high enough electric field to produce an arc on its own inside the cavity. To produce an arc, a means of concentrating the electric field must exist. In metal, electrons or electric charges move freely and are easily separated. If metal is introduced into a microwave oven, the cavity electric field causes the electrons to move in the metal. The electrons tend to concentrate at sharp points or edges. The charge in this region further concentrates the electric field. If a second piece of metal is introduced into the oven, or if the first piece is moved close to the cavity metal wall, further concentration of the field occurs. According to Equation 1-1, the field is produced when charge is separated, causing a potential or voltage V to be developed. If the distance d between the separated charges is reduced, the field is increased or concentrated.

Thus, as the metal-to-metal distance decreases, the field increases. For distances around 0.001–0.002 in (0.03–0.05 mm), arcing could easily occur. Arcing can be prevented if metal is not placed next to metal. Metal dishes, trays, and lidstock should be separated from cavity walls and adjacent metallic packaging. It is comforting to know that if an arc does occur with this type of packaging, it will mostly startle the consumer. Other than a small blemish on the container or in the cavity paint, little damage will occur. The arc quickly extinguishes itself as the metal in the microscopic region of the arc melts and the distance between separated charge increases. Twist ties, on the other hand, must not be used since they are usually wrapped with paper and will burn if an arc occurs.

Pottery, china, and glasses with decorative bands or trim are a major problem if used in the microwave oven. The decoration is usually applied with metallized paint, the metal particles of which are separated by microscopic gaps. Placing such objects in a microwave oven usually results in a spectacular flash because all the particles arc at once, and the aesthetic quality of the trim is completely destroyed.

Products with high salt contents, such as prepared meats, hot dogs, and cheese, might have a high enough conductivity to arc when placed adjacent to each other or near metal container or wall. Foods with high water content and cut with sharp edges, such as raw julienne vegetables, could also arc.

PRODUCT SAFETY

The concern over the safety of a product heated or processed by microwaves falls into four categories: two of no major significance—residual radiation and nutrition—one of minimal significance—chemical migration—and one in which great care and concern must be exercised—pathogen growth.

Residual Radiation

Since microwaves are a form of electromagnetic radiation, they have erroneously been linked to nuclear and ionizing radiation. Hence, some of the fear associated with residual radiation in high-energy or nuclear-irradiated food products has spilled over into the microwave field. As was previously discussed, since microwaves are low-photon-energy, nonionizing radiation, they produce *only* heating effects, and do not induce residual radiation in foods. Microwaves themselves are not retained in the food after it is removed from the microwave oven. The continued cooking that occurs during the "standing time" following heating is due to the redistribution of the thermal energy, or heat, by conduction within the food (Buffler 1991).

Nutrition

There is concern that microwave radiation could interact negatively with food nutrients. Again, because microwaves are low-energy radiation, they can only heat the food products. There is no theoretical basis for any other mechanism nor experimental evidence to the contrary that indicates a direct, nonthermal interaction with microwaves.*

Much research on the nutritive retention of microwaved vegetables, meats, and bread has shown that microwave cooking actually enhances the nutritive content of food, compared with the conventional process. During microwaving, less leaching of vitamins and minerals occurs due to the shorter heating time and smaller amount of water used. Other researchers have found no difference between microwave and conventional preparation, while virtually no reports exist for microwaved foods with less nutritive value than conventional. Papers presenting the research on the nutritional aspects of microwave cooking are included in a microwave research bibliography (IMPI 1984).

Chemical Migration

Because of the intrinsic nonbrowning property of the microwave oven, susceptor heater boards (Chapter 6) have been used in microwave packaging to brown and crisp product surfaces. These susceptors are usually constructed of a metallized polyethylene terephthalate (PET) film laminated to a paperboard backing with an adhesive (Chapter 7). The susceptors reach temperatures near 230°C (480°F) (Lentz and Crossett 1988; Kashtock, Wurts, and Hamlin 1990) and may allow the migration of either unwanted plasticizers from the adhesive or oligomers from the plastic film to the heated food product. Since PET is a nonplasticized film, no plasticizer migration occurs from the film itself.

Initially it was thought that the PET film would act as a barrier, keeping the adhesive plasticizers from contaminating food surfaces. Recent work, however, has shown that this is not so (Begley and Hollifield 1990). Using ultraviolet absorption techniques, they determined that plasticizers from the adhesive migrate through the film and into a food simulant. Also

---

*Many cases of direct, nonthermal chemical reactions or enhancements due to microwaves have been reported in the literature. In all thoroughly investigated cases, the reported results were due to a misunderstanding of the thermal heating profile of the experiment. A recent case reporting an athermal-enhanced chemical reaction has been analyzed (Jahngen et al. 1990) and shown to be strictly thermal in nature.

oligomers from the film itself were found to migrate into the food simulants.

Under FDA regulations (FDA 1958), susceptors are not directly covered. Manufacturers of materials that do make food contact, such as susceptors, are responsible for the safety of their product. To investigate the safety issue, the FDA gathered data on 42 commercial susceptors from 15 companies (FDA 1989) and found that 5 showed higher than acceptable levels for only one toxin, benzene. The suppliers of these susceptors are now reducing contaminants to acceptable levels by modifying manufacturing techniques.

The American Society of Testing and Materials (ASTM) is preparing a test procedure for determining volatile and nonvolatile migrants from susceptors to food simulants (ASTM 1991). The FDA is continuing to evaluate data and will undoubtedly codify testing procedures and limits of migrants. In the meantime they have not felt that the hazard involved in susceptor use mandates removing them from the marketplace. A position statement by the FDA is expected by 1992. The issue has been reviewed by Borodinsky (1988).

A second migration concern is the leaching of plasticizers from plastic film or cling wrap placed over food heated in a microwave oven. The concern is highest for fatty foods where temperatures could be 450°F in desiccated regions of the food. If the film contacts these regions, unacceptable migration into the food has been reported for certain chemical compositions of film (MAFF 1987). In the United Kingdom, a statement has been published: "Government advice is that neither PVC nor polyolefin cling films should be used in intimate contact with food during cooking in a microwave oven." (MAFF 1990) If caution is used in maintaining product temperatures below 93°C (200°F) and keeping the film away from the food, no problems should be encountered.

Pathogen Growth and Survival in Microwaved Foods

Illness from food, often called food poisoning, arises from three general sources. First, living pathogenic or illness-causing microorganisms can contaminate a food product. When living, these pathogens are called vegetative. These living microorganisms produce toxins that, if ingested, can cause serious illness. Second, in foods, the pathogens might change their state into the form of spores, which might lie dormant until environmental conditions are appropriate for their activation. When activated, toxins may be formed directly or new microorganisms may be produced. Third, pathogens can produce toxins directly in the food that remain until ingestion even though the pathogens themselves may have died.

Consumers usually assume that the food they buy is free from harmful contamination, and they simply need to properly prepare it to avoid health problems. Strictly speaking, the only foods free of illness-producing microbes and toxins are commercially sterilized products that have been canned or retorted at 121°C (250°F) to achieve a 10-fold reduction of *Clostridium botulinum* spores. *C. botulinum* is the most heat-resistant and most dangerous pathogenic microorganism in foods and is thus used as the standard for sterilization. The only way to ensure that all types of foods and food products are free of pathogens, toxins, and poisons is for the supplier and the consumer to process and handle the food in a way that prevents microorganism contamination and toxin production.

Pasteurized food products are those in which the vegetative pathogenic microbes are inactivated. Milk pasteurization at 63°C (145°F) for 30 min or 72°C (162°F) for 15 s are typical process standards. Pasteurizing does not destroy spores, however, so pasteurized foods must be kept below 3°C (38°F) to ensure safety. The quality of pasteurized foods decreases during storage because the spoilage microorganisms produce by-products, such as enzymes, that alter odor, flavor, and texture. Pasteurized products have a remarkable history of safety even when temperature-abused because they usually spoil and become inedible before they become dangerous to health.

Foods from chilled cases such as meats and poultry as well as delicatessen-prepared products might contain pathogens, depending on the contamination and preparation they have undergone.

Contaminating organisms could enter the food during growth, slaughtering, harvesting, processing, packing, transportation, storage, or preparation. Unwanted pathogens, if allowed to multiply to hazardous levels, cause, at the least, nausea, vomiting, diarrhea, and, possibly, death if appropriate controls are not maintained throughout the entire food supply chain.

Types of Pathogens

Pathogenic microorganisms include bacteria, their reproductive organisms known as spores, viruses, molds, and parasites. A very common infective food bacterium is salmonella (*Salmonella enteritidis* and other species) found in raw meat, poultry, fish, and dairy products. *Listeria monocytogenes,* another pathogen, has been isolated from meat, poultry, and dairy products. The ubiquitous *Staphylococcus aureus* is found not only in animal tissue but also is carried by humans, who often contaminate the food during preparation. The toxin from this bacterium is not easily inactivated, being able to survive boiling-water temperatures. It is therefore critical to prevent its formation.

Another common disease-causing organism is *Trichinella spiralis,* a parasitic nematode found in pork. The eggs of ingested *T. spiralis* can hatch in human muscle tissue, causing severe illness and even death. Fortunately, good cooking practice in the United States and carcass inspection in other countries have made this disease rare.

Foodborne illness from poorly refrigerated fish can be caused by decomposition products such as histamines. Viruses such as *Hepatitis A* or *Norwalk* can also be implicated. In shellfish, microbial contamination from bacteria such as *Vibrio vulnificus* and other bacterial species can be responsible for illness. On occasion, when fish ingest certain plankton, such as the well-known red tide found on the East Coast of the United States, a very toxic poison is produced. This toxin remains in the fish muscle and can cause severe illness if consumed.

Aspergillus flavus is a mold that occasionally contaminates grains, nuts, and pulses (e.g., dried beans, peas) that are not harvested and stored properly. This mold produces aflatoxin, a very potent liver cancer agent. The United States government has set safe limits for the allowable levels of this toxin in products.

Preparation and Handling

The key defense against foodborne illness is proper handling of food during all phases of its production chain. Care must be taken to avoid contamination and microorganism multiplication during growth and transportation of raw materials. Human handling must be done under strictly controlled manufacturing practices. Strict attention should be paid to processing and storage temperatures and times appropriate to each product.

Finally, to make sure that all remaining foodborne pathogens have been inactivated, safe cooking procedures have been developed for most types of edible products. Because all cooking appliances, including microwave ovens, might not heat some products uniformly, it is important that the coldest spot of a heated or cooked product reach the recommended cooking temperature for the appropriate amount of time. Because of the shorter cooking time usually associated with the microwave oven and the possibility for cold spots to exist, specific procedures have been recommended for microwave heating and cooking. Many are prudent for all types of food preparation:

1. Food preparers should always wash their hands before handling foods, including cleaning fingertips and under fingernails.
2. Do not cross-contaminate the food preparation area. Cutting boards and chopping blocks should be thoroughly cleaned and sanitized after

cutting meats and before being used for other products. Do not cross-contaminate by wiping utensils on sponges. Wash utensils thoroughly before reusing.

3. Do not leave refrigerated food at room temperature for more than 2 h before cooking.

4. Cook pork in a closed container such as a loosely sealed cooking bag or a covered microwave safe container. A final temperature of 77°C (170°F) is considered safe. Whole roasts should be covered in the bag with foil and allowed to stand 10 min before serving; no standing time is required for smaller pieces such as ribs or chops (NLS&MB 1984).

5. Cover whole chicken with wax paper and cook breast side down on medium for 15 min, turn over and cook on medium for 15 min more. Allow to stand covered for 5 min before serving. It is strongly recommended to purchase a meat thermometer to check doneness of 71°C (160°F) after standing. Cover chicken parts with wax paper and cook on high for 6 min per pound. Turn dish 45° halfway through cooking (NBC).

6. Hot foods must be cooled to 4.5°C (40°F) within 4 h to ensure safety. Typically a hot dish needs to be refrigerated within 2 h of cooking to meet this requirement.

7. Many refrigerated foods, especially leftovers, should be used within five days to prevent microorganisms from multiplying to illness-causing levels.

8. Reheat leftovers, preferably, in a covered microwave-safe container. If covering with plastic wrap, attempt to tent the wrap above the food. Do not allow the wrap to melt on food.

Training

Ensuring the safety of food requires knowledge and attention to correct procedures. Any person working professionally in the field of food service should have specific training in the appropriate microbiological safety issues (Snyder 1990). A detailed treatment of food safety issues appropriate for the food service personnel or the consumer is covered in a publication by the U.S. Department of Agriculture (USDA 1990a). More consumer-oriented information is also available (USDA 1990b; FMI 1991). For specific questions on meat and poultry, the U.S. Department of Agriculture has established a hotline, 800–535–4555.

References

ANSI 1982. American National Standards Institute. *Safety Levels with Respect to Human Exposure to Radio Frequency Electromagnetic Fields, 300 kHz to 100 GHz*. New York: Institute of Electrical and Electronic Engineers, Inc.

ASTM 1991. Private commuication. American Society for Testing and Materials, 1916 Race Street, Philadelphia, PA.

Begley, T. H. and Hollifield, H. C. 1990. Migration of dibenzoate plasticizers and polyethylene terephthalate cycle oligomers from microwave susceptor packaging into food-simulating liquids and food. *Journal of Food Protection* **53**(12):1062–1066.

Borodinsky, L. 1988. FDA perspectives on microwavable packaging materials & food/package interactions. Paper read at MW Foods, 8–9 March 1988, Chicago, IL. (Available through Food and Drug Administration, Center for Food Safety and Nutrition, Washington, DC 20204)

Braithwaite, L., Morrison, W., Otten, L., Pei, D., McMillan, I. 1989. A technique for rewarming hypothermic piglets using microwaves. Paper read at Canadian Society of Animal Science Meeting, 9–13 July 1989, at Montreal, Quebec.

Braithwaite, L., Morrison, W., Bate, L., Otten, L., Hunter, B., and Pei, D., 1991. Effect of exposure to operant-controlled microwaves on certain blood and immuniological parameters in the young chick. *Poultry Science* **70**:509–514.

Buffler, C. 1988. Whole body microwave heating of humans and livestock. *Harvard Graduate Society Newsletter,* Summer 1988, p. 11–13.

Buffler, C. 1991. Perceived rapid cooling of microwaved foods. *Microwave World* **12**(2):16–18.

Buffler, C. and Lindstrom, T. 1988. Experimental evidence of water eruption caused by super-heating. *Microwave World* **9**(4):10–11.

Cheung, A. and Al-Atrash, J. 1987. Microwave hyperthermia for cancer therapy. *Proceedings of the IEE* **134A**(6):493–522.

Decareau, R. 1978. Evaluation of aluminum foil container performance in microwave ovens. Proceedings of the 13th Annual Symposium of the International Microwave Power Institute, pp. 12–13.

EMC 1991. EMC Technology. Star Route 625, Gainsville, VA.

FCC 1991*a*. Federal Communications Commission. "Industrial, Scientific and Medical Equipment," *Code of Federal Regulations,* Title 47, Part 18.

FCC 1991*b*. Federal Communications Commission. "Radio Frequency Devices," *Code of Federal Regulations,* Title 47, Part 15.

FDA 1958. Food and Drug Administration. Federal Food, Drug and Cosmetic Act, as amended. Title 21. *Code of Federal Regulations,* Section 201.

FDA 1989. *The Determination of Volatile Substances From Commercial and Developmental Susceptor Microwave Food Packaging.* The Susceptor Microwave Packaging Committee. (Submitted to Food and Drug Administration, Division of Food and Color Additives, HFF-335, Washington, DC.)

FDA 1991. Food and Drug Administration, Center for Devices and Radiological Health. Regulations for the Administration and Enforcement of the Radiation Control, Health and Safety Act of 1968. *Code of Federal Regulations,* Title 21, Part 1000 to 1030.

FMI 1991. *Food Safety and the Microwave.* Washington, DC: Food Marketing Institute.

Ghandi, O. 1990. *Biological Effects in Medical Applications of Electromagnetic Energy.* New York: Prentice-Hall.

Guy, A. and Chou, C-K, 1983. Electromagnetic heating for therapy. In *Microwaves and Thermoregulation* (E. Adair, ed.), pp. 57–93. New York: Academic Press.

HEW 1979. *Inexpensive Microwave Survey Instruments,* U.S. Department of Health, Education and Welfare. HEW Publication (FDA) 80-8102.

IMPI 1984. *Microwave Research Bibliography; 1970-1983.* Clifton, VA: International Microwave Power Institute.

Jahngen, G., Lentz, R., Pesheck, P., and Sackett, P. 1990. Hydrolysis of adenosine triphosphate by conventional or microwave heating. *Journal of Organic Chemistry* **55**:3406–3409.

Kashtock, M., Wurts, C., and Hamlin, R. 1990. A multilab study of food/susceptor interface temperatures measured during microwave preparation of commercial food products. *Journal of Packaging Technology.* **March:**14–19.

Lentz, R. and Crossett, T. 1988. Food/susceptor interface temperature during microwave heating. *Microwave World* **9**(5):11–16.

Litton 1971. *An Exciting New World of Microwave Cooking.* Minneapolis, MN: Litton Microwave Cooking Products (now doing business as Menumaster, Sioux Falls, SD).

MAFF 1987. *Survey of Plasticiser Levels in Food Contact Materials in Foods.* Food Surveillance Paper No. 21. Ministry of Agriculture, Fisheries and Food. London: HMSO Publications.

MAFF 1990. *Plasticisers: Contintinuing Surveillance.* Food Surveillance Paper No. 30. Ministry of Agriculture, Fisheries and Food. London: HMSO Publications.

Medtronics. Letter from Technical Consultant. Medtronics, Inc., Minneapolis, MN.

Metaxas, R. 1988. Towards a unified approach to teaching electroheat: direct resistance to microwave heating and laser to plasma jet. *Electromagnetic Energy Review* **1**:33–38.

NBC. *Chicken Microwaves in Minutes.* Columbia, SC: National Broiler Council.

NLS&MB 1984. *Easy Steps: Microwave Cooking With Pork,* Chicago, IL: National Live Stock & Meat Board.

Osepchuk, J. 1973. *Hearings Before the Committee on Commerce; U.S. Senate on Public Law 90-602.* Publication 93-24, p. 118. Washington, DC: U.S. Government Printing Office.

Osepchuk, J. 1978. A review of microwave oven safety. *Journal of Microwave Power* **13**(1):13–26.

Osepchuk, J. 1981. Debunking a mythical hazard. *Microwave World* **2**(6):16–19.

Osepchuk, J. 1984. A history of microwave heating applications. *IEEE Transactions on Microwave Theory and Techniques* **32**(9):1200–1224.

OSHA 1990. *General Industrial Standards.* Occupational Safety and Health Administration. *Code of Federal Regulations,* Title 29, Part 1900.

Packaging Strategies 1989. *The 1990's . . . The Microwave Decade.* Westchester, PA.

Raytheon 1982. *Use of Metal in Microwave Ovens.* Bulletin MM4. Raytheon Company, New Products Center, Burlington, MA.

Savin, B. 1983. New trends in the standardization of microwave electromagnetic radiation. Trudi Professionalnikm Zabolevanii. **3**:1–4.

Snyder O. 1990. *Food Safety through Quality Assurance Management.* St. Paul, MN: Hospitality Institute of Technology and Management.

USDA 1990*a*. *Preventing Foodborne Illness: A Guide to Safe Food Handling.* U.S. Department of Agriculture Food Safety and Inspection Service. Home and Garden Bulletin No. 247.

USDA 1990*b*. *A Quick Consumer Guide to Safe Food Handling.* U.S. Department of Agriculture Food Safety and Inspection Service. Home and Garden Bulletin No. 248.

Note: Most United States government publications referenced are available through the U.S. Government Printing Office, Washington, DC.

10

Microwave Processing of Foods

For the last 25 years expectations have been high that radio frequency (RF) and microwave processing of foods might find a niche in the industry. Unfortunately there has been only modest growth in sales of microwave processing equipment over this period. Numerous food processing applications, which became popular in years past, such as chicken cooking, potato chip processing, and doughnut proofing and frying, have virtually all been replaced by conventional techniques. Only tempering, bacon cooking, and pasta drying have retained a major foothold in the market. Reasons for the lack of success of microwave processing in the food industry are many. Newness and unfamiliarity with microwave equipment always presents an initial obstacle. Economics, including cost of capital equipment, invariably plays a role in any final procurement decision. An excellent review of the history of microwave processing has been published by Osepchuk (1984). O'Meara (1973) has given a review of the problems surrounding the application of microwaves to the food processing industry. His analysis is as valid today as it was in 1973. A survey of current microwave industrial usge is given by Smith (1984).

The technology of microwave processing has been amply covered by Copson (1962, 1975), Püschner (1966), and Metaxas and Meredith (1983). A wide variety of industrial applications, including food processing, have also been covered (Decareau 1985, 1990; Mudgett 1982; Decareau and Peterson 1986; Ohlsson 1989; Stuchly and Stuchly 1983). These latter references should be reviewed by all contemplating the use of microwave power for food processing. Due to the extensive coverage in the literature, this chapter is brief. A broad overview of the possible applications of microwave energy to the food industry is given. Simple guidelines for the food producer to use to determine if microwave processing may provide a viable option for a given product are presented.

PROCESSING AND DRYING

The use of electromagnetic energy for heating and processing foods must have been considered at the Radiation Laboratory at MIT during World War II. The technology of radar, developed at this laboratory, was experimented with during the 1950s for food processing. A continuous microwave processing system, using a conveyor for product transport, was introduced by the Philips Co. of Eindhoven, Netherlands, in 1960. Following that, with Cryodry's introduction of a conveyorized system in 1962, microwave processing began in earnest. Since that time many applications to the food industry have been explored, some with temporary, some with lasting, success. Table 10–1 lists applications and comments as to their effectiveness. A list of the RF and microwave frequencies allocated by the Federal Communications Commission (FCC) for industrial (as well as scientific and medical) applications is presented in Table 10–2.

There are several reasons for the failure of microwave food processing to develop into a full-fledged industry. The most prominent are as follows, in order of importance.

1. Economics.   Microwave processing is viewed as an efficient process because of its association with the "efficient" microwave oven. In actuality, both microwave cooking and microwave processing are approximately 50% efficient when the power delivered to the load is compared with that drawn from the power line. Economics, particularly when compared with gas, allows microwave processing to compete only under special circumstances.

2. Newness of Technology.   Even though microwave equipment has been in use industrially for 30 years, it is still deemed new and high tech by plant managers who have been processing by conventional techniques for decades.

3. Safety.   Microwave energy has unfortunately received a bad reputation because its name has been associated with nuclear radiation. Even though safety is not an issue with microwave equipment, its use is still resisted by many plant employees.

Microwave Drying and Dehydration

Table 10–1 lists several examples of microwave drying. Specifically, pasta and potato chips have been dried successfully. Freeze drying and vacuum drying, in conjunction with microwave energy, have also shown promise.

TABLE 10-1   Applications of Microwaves to the Food Processing Industry*

Application	Remarks
Bacon cooking	Successful application. Used to precook, primarily for fast food chains.
Bread baking	Few, if any systems in operation. Advantage in space saving when combined with proofing for bread to be finish-browned by consumer. May have application when combined with convection or infrared heat.
Chicken cooking	Early success in cutting labor and improving quality of product. No longer used. Reason unknown.
Doughnut processing	Early success in combining microwave proofing and microwave-assisted frying. Presently not used.
Pasta drying	Successful application. Industry dominated by Microdry, Inc. Golden Grain Co. largest user.
Potato chip finish drying	First large-scale use of microwaves for food processing. Prevented browning on potatoes with wide range of sugar content. Became obsolete when raw material specifications tightened and transportation time to plant decreased.
Sausage cooking	Limited present use. Reduces drip (moisture) loss appreciably during processing.
Blanching	Quality improvement, but not economical for commodity products. May have excellent application to expensive, sensitive foods.
Freeze drying	In experimental use only. No economical consumer application found to date.
Tempering	Major success of microwave processing for tempering large blocks of frozen foods. Food is heated to just under freezing temperatures, allowing easy chopping, cutting, processing, etc. Industry dominated by Raytheon Co.
Vacuum drying	Previously used for producing juice concentrates. Provides higher retention of volatiles than conventional processes. Expensive. No large-scale use.
Thawing	Major challenge for industry is to overcome runaway heating at corners and edges. Present applications temper first, then allow to finish thawing at ambient temperature.

*Source: Information extracted from Decareau (1985) and Ohlsson (1989).

The lowering of the boiling point by reducing pressure (cf. Table 10-3) allows sensitive products to be dried at lower temperatures. It is critical to note that the above applications are all unique. They all provide a product that is potentially superior in quality to the product produced by conventional techniques. This point is key to almost all industrial microwave processes. If one compares the use of microwave drying with conventional

TABLE 10-2   ISM Frequency Allocation for Electromagnetic Heating and Processing in the United States*

Frequency	Bandwidth (± MHz)	Typical Use (Zimmerly 1991, Wilson 1987)
6.78 MHz	0.015	Plastic preheating, wood drying and bonding
13.56 MHz	0.007	Plastic preheating, wood drying and bonding, textile and miscellaneous drying and curing, ceramics
27.12 MHz	0.163	Plastic preheating, textile drying and curing, paper product bonding, plastic welding
40.68 MHz	0.02	Plastic preheating, paper product bonding
915 MHz	13	Microwave processing and tempering; no commercial or consumer ovens
2,450 MHz	50	Microwave ovens, microwave processing
5.8 GHz	75	No active usage
24.125 GHz	125	No active usage
61.25 GHz	250	No active usage
122.5 GHz	500	No active usage
245 GHz	1,000	No active usage

*Note: 896 MHz is allocated for use in the United Kingdom but not in Europe. A number of manufacturers, however, have had 915-MHz equipment accepted for use in Europe by keeping interference emission below acceptable levels for the country of installation.
Source: FCC (1991).

process, one can readily see that microwave drying by the removal of bulk water is not economically viable. Figure 10-1 shows a typical drying (or water loss) curve of a product undergoing conventional processing. The curve has three regions. First, the increasing-rate region occurs when the product is initially exposed to the process. The product temperature increases to the boiling temperature of the water contained inside it. Second, the constant-rate region is where equilibrium is reached and moisture is driven off at a constant rate. Third, when a large portion of the moisture has evaporated, a decreasing rate sets in. This occurs when the internal moisture transport is limited by its ability to diffuse through the partially dried product.

Microwave drying has its major advantage in speeding up the decreasing-rate "tail" of the curve. In many materials this region of the drying process is a major portion of the process time. For some applications decreasing the heat-up time may also offer an advantage. It is highly unlikely, however, that an economic advantage will be demonstrated if only bulk water removal is desired, such as occurs in the constant-rate region. The increased drying speed using a microwave-augmented system is shown as the second curve in Figure 10-1. Here a three-zoned system could be used. First, the product is fed into a short microwave system where the product is brought

TABLE 10-3   Boiling Temperature of Water as a Function of Pressure

pascals (absolute)	torr (mm Hg)	psia	psig	atm (bar)*	°C	°F	Note
1,227	9.2	0.18	—	~0.01	10	50	
3,173	23.8	0.46	—	~0.03	25	77	1
12,332	92.5	1.79	—	~0.12	50	122	
50,662	380	7.35	—	0.5	82	180	2
83,560	627	12.1	—	0.825	94.7	202.5	3
101,325	760	14.7	0	1	100	212	
202,650	1520	29.4	14.7	2	120.6	249	
205,583	1542	29.8	15.1	~2	121.1	250	4
506,625	3800	73.5	58.8	5	152.3	306.2	

psia = pounds per square inch, absolute.

psig = pounds per square inch, gauge (As measured on a pressure gauge, i.e., pressure above atmospheric pressure)

atm = atmospheres or approximately bars.* Pressure normalized to standard atmospheric pressure defined as 960 mm of mercury.)

Notes

1. Severe microwave plasma discharge occurs below this pressure.
2. Approximate pressure inside "pressurized" aircraft.
3. Barometric pressure at Denver, CO (altitude = 5282 ft = 1610 m).
4. Common sterilization temperature.

*One bar does not exactly equal 1 atm. 1 bar = 10^5 pascals, 1 atm = 101,325 pascals. The use of the unit *bar* is discouraged.

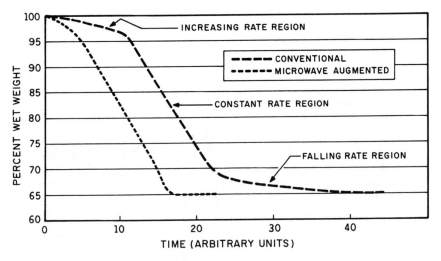

FIGURE 10-1.   Idealized Product Drying Curve.

up to temperature. It is then passed into a conventional drying system where bulk water is removed and, finally, into a second microwave tunnel where the final drying takes place.

It is important to understand the dielectric properties of the material being processed, especially the properties of the dried material. If the loss factor of the dried product is high, such as with many materials containing cellulose, continued microwave heating occurs, with the possibility of thermal runaway. If the loss factor is low, portions dried first no longer heat, whereas moisture-containing portions continue to do so. This "leveling" effect affords a major advantage for even drying of these types of materials.

Pasta drying is a perfect example of how only a small amount of microwave power can be of large benefit to the process. Pasta must be dried to a very low moisture level. Unfortunately the diffusion rate through the partially dried pasta is low, presenting a very long decreasing-rate region. The only way to speed the process is to increase the ambient drying temperature. This procedure "case-hardens" or cracks the outside, creating an unacceptable product. Since microwaves penetrate the product, they produce a pressure gradient that pumps out the moisture. This property can be used to advantage to speed up the drying process and produce a superior product at the same time.

If process time can be decreased by replacing or augmenting a conventional line with a microwave line, throughput may be increased by replacing old equipment with new, while utilizing the same amount of floor space. It might thus be more economical to obtain increased production by installing a microwave line than by installing a larger conventional system. This especially holds if space is limited or if expanded throughput could only be obtained by moving into a new building.

PASTEURIZATION AND STERILIZATION

Pasteurization

Pasteurization of foods by microwave processing has been successfully accomplished for decades, especially in Europe. The process is particularly suited for bread products, plated meals, and soft cheeses. The major advantage of the microwave process is that the product may be pasteurized after it has been sealed in its package. Shelf life thus can be extended from days to over a month without preservatives. A status report on pasteurization was given by Decareau (1988). A typical microwave pasteurization tunnel is illustrated in Figure 10–2. Manufacturers of pasteurization equipment are listed in Appendix 4.

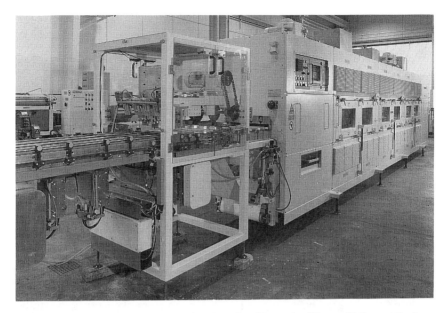

FIGURE 10-2.   Microwave Pasteurizing Tunnel   (Courtesy of Berstorff Corporation).

Sterilization

To date, microwave sterilization has gained little of the acceptance that pasteurization has realized, even though microwave-sterilized, room-temperature, shelf-stable meals, for example, might have a major potential advantage over their retorted counterparts. Because the heat-up time of the microwave process is much faster than the heat-up time in a retort, the product's organoleptic qualities could be considerably improved. Little literature comparing qualities of microwave with conventionally processed foods is available. The available literature indicates microwave-processed food has a higher quality rating than retorted foods and only a slightly lower rating than frozen foods (Ohlsson 1987).

There are two major reasons for the slow acceptance of microwave sterilization: economics and, most important, microbiological safety.

Economic consideration arises because microwave sterilization is costly. Several European manufacturers have equipment presently under development or for sale. Pressures up to 45 psig are required during the microwave process, which considerably complicates equipment design. Also, the well-known problem of providing a uniformly heated product makes it extremely time-consuming and costly to adjust the microwave pattern to produce the quality advantage theoretically possible by using microwaves.

Often, each product could require custom adjustment. These costs could make microwave sterilization too costly to justify the quality improvements obtained.

The nonuniformity problem previously discussed may have an even more serious impact on the second consideration, safety. The ability of the deadly *Clostridium botulinum* to grow readily in nonacidic (pH > 4.6) anaerobic (oxygen-free) environments has been a major concern to the canning industry for decades. It is thus imperative that all portions of a sterilized product be brought to sufficient temperature for a sufficient time to ensure that all *C. botulinum* spores have been killed. It is well accepted that a temperature of 250°F (121°C) for 3 min is sufficient to accomplish such a kill. The time is usually doubled to 6 min to provide a safety factor. To reach these temperatures without boiling requires pressurization of the microwave system. The boiling temperature of water as a function of pressure is given in Table 10–3. An overview of conventional thermal processing and its microbiological aspects is given by Earle (1983).

The presence of hot and cold spots caused by uneven heating makes it very difficult to ensure that all portions of a meal have reached a kill temperature. In addition, the technical ability to accurately measure the temperature distribution throughout an entire microwave sterilized product has not yet been demonstrated. The most reliable, albeit statistical, method to ensure safety is to sample products from the production line and subject them to intensive bacteriological testing. Correlation of bacteriological tests with fiber-optic temperature probes, infrared thermography, and/or inserted data recorders (Appendix 4) could eventually lead to sufficient confidence to place a product on the U.S. market.

The United States Department of Agriculture (USDA) has not yet approved the use of microwave processing for food sterilization. To obtain such an approval for food for human consumption will most certainly be an arduous task. In Europe, microwave-sterilized foods, primarily pasta dishes such as lasagna and ravioli, are on many grocery shelves with no reported difficulties. These products are room-temperature shelf-stable with a shelf-life date code of one year. Safety regulations are less stringent in Europe, allowing these products early access to the market. The success of European microwave-sterilized products will be watched carefully by U.S. manufacturers.

ANALYSIS OF PROCESSING ECONOMICS

Chapter 6 provides the basis for understanding food heating in a microwave process. The laws of thermodynamics apply to microwave as well as conventional processing. Many persons considering the use of microwaves for their process contemplate purchasing or leasing a high-power microwave system to perform preliminary experiments. From a practical point of view, it is much wiser to spend a short amount of time, and minimal effort to determine if there is at least a chance that a microwave process may be economical. The first step is to perform a simple calculation to determine the minimum theoretical power required to perform the process. Second, a simple experiment in a home or commercial microwave oven will provide the maximum power required. Between these two values is the actual power needed in production. If the value derived from these two steps appears economically promising, the next step is to contract for prototype experimentation. The following parameters must be known to perform these steps.

1. Throughput. The first consideration in a process evaluation is to determine production throughput, or the number of pounds of product per hour required from a production line during operation. Careful consideration should be given to the number of shifts for which the equipment is operated: the more shifts, the higher the daily throughput. Cleaning and maintenance time must also be planned and subtracted from production time.

For ease of calculation, the raw material input rate R is used. This is the actual production throughput, increased by the amount of moisture lost during processing.

2. Temperature Rise. During processing most foods are brought to the boiling temperature of the water within them. The exception is the process of sterilization, where higher temperatures are required under pressure. For an exact calculation, the initial temperature of the food should be determined so that the total rise in temperature, ΔT, can be determined.

3. Moisture Loss. During processing most foods lose moisture. This loss indicates that the product has been heated to the boiling temperature of water, causing evaporation to take place. From the conventional process, the fraction moisture lost, f, during the process must be determined. Thus, a product whose weight decreased from 1.25 to 0.95 lb would have a weight loss fraction $f = 0.24$ due to evaporation.

4. Heat Capacity. The heat capacity, C_p, determines the amount of energy required to raise the temperature of the product a given amount. Thermal data for a wide variety of foods is found in the literature (see Chapter 6). If data for the product cannot be found, it can be easily calculated if its analysis into water, carbohydrate, protein, fat and ash is known. The heat capacity is then given by Equations 6–2a.

5. Heat of Vaporization (latent heat). A fixed amount of energy is required to convert any liquid to a gas or vapor. For foods, the liquid of importance is water. It takes 970 Btu/lb of energy to evaporate water (539 kcal/kg, 2.257 kJ/g) at 100°C (212°F) and 1 atm of pressure. Vaporization occurs at other temperatures, depending on the pressure (Table 10-3).

Theoretical Considerations

With the previous five parameters known, the theoretical amount of power required to process the product can be determined. This calculated power is a best-case value and represents a nonobtainable 100% theoretical process efficiency. The theoretical value is important to determine since it indicates the lowest amount of power required to meet the throughput requirement.

The power requirement calculation can be broken down into two parts. First, the power P, in kilowatts, to raise the product to the boiling point is

$$P \text{ (kW)} = \frac{R \text{ (lb/h)} \times C_p \text{ (Btu/lb°F)} \times \Delta T \text{ (°F)}}{3414 \text{ (Btu/kWh)}} \qquad (10\text{-}1)$$

The equations are given in English units. Conversion to metric is easily accomplished, especially with conversion software programs such as SI PLUS (1989).

Second, the power to evaporate the moisture is determined by

$$P \text{ (kW)} = \frac{f \times R \text{ (lb/h)} \times 970 \text{ (Btu/lb)}}{3414 \text{ (Btu/kWh)}} \qquad (10\text{-}2)$$

The total power requirement for the process is the sum of Equations 10-1 and 10-2.

A very useful term, process efficiency E_{ff}, can be defined by dividing the raw material input by the total required power:

$$E_{ff} \text{ (lb/kWh)} = \frac{R \text{ (lb/h)}}{P \text{ (kW)}} \qquad (10\text{-}3)$$

Note that E_{ff} can be thought of either in terms of pounds per kilowatt hour or its equivalent, pounds per hour per kilowatt.

The total theoretical power can be approximated for practical estimation purposes by the following technique. If one assumes a worst-case situation for Equation 10-1, where the product is all water, $C_p = 1$ Btu/lb (4.186 J/g°C), a starting temperature of 70° F (21°C), and $R = 1$ lb/h (2.2 kg/h), one finds that 0.326 kWh (1.17×10^3 kJ) is required to raise the product to boiling and evaporate 1 lb (0.454 kg) of water. For this process E_{ff} is the reciprocal of this value, or approximately 3 lb/kWh (1.83×10^{-3} kg/kJ).

Dividing this simple number into the amount of water evaporated during the process, f × R allows one to calculate quite easily the theoretical microwave power required for a process. This value is the minimum power required. If this power calculates to be too high or too costly, no more investigation is required.

Practical Considerations

If the theoretical numbers appear within reason, a simple test in a home or commercial oven should be tried. Place a sample of the product to be processed in the microwave oven. A sample size of approximately 1 lb (0.454 kg) is appropriate. Heat the sample until it reaches the appropriate process state required of the production line. Make sure the weight loss or water loss is similar to that found in the conventional process. To calculate a process efficiency the oven's power must be known. Since the calculation will be very rough, the power measurement technique is not important. The manufacturer's labeled power is sufficient. If this power is not available, the 2-L test of Appendix 5 can be used. The process efficiency is

$$E_{ff} \text{ (lb/kWh)} = \frac{W \text{ (lb)}}{P \text{ (kW)} \times t \text{ (h)}} \tag{10-4}$$

where W =   sample weight before heating

   P =   oven power converted to kilowatts,

   t =   heating time converted to hours.

Using the required raw material input rate R with Equation 10–3, one can estimate a worst-case or maximum power required: P (kW) = R (lb/h)/E_{ff} (lb/kWh).

The worst-case power is seldom more than 75% higher than the theoretical power. Microwave processing lines are typically 60–80% efficient in converting their generator power to power absorbed by the load. Thus, one may be quite confident that a power between the theoretical power and the worst-case power will adequately provide the required process throughput. Recall that microwave systems are only 50% efficient in converting line power to microwave power. For electrical usage calculations one must double the calculated microwave power.

Once the estimated power is determined, capital equipment costs can be estimated. Before determining costs, a trade-off analysis must be made as to which microwave frequency should be used. Processing equipment is available at both 2450 and 915 MHz. A comparison of the two frequencies is given in Table 10-4.

Rough capital costs can be obtained from Table 10-4 after obtaining the estimated microwave power from previous calculations. Because of the large size of 915-MHz generators and their corresponding lower cost per kilowatt, they are usually preferred by most large-scale users, particularly if throughput requirements are large. Tempering, for example, is primarily done at 915 MHz in the United States, with some 2450-MHz systems being installed in Japan and Europe. On the other hand, the adaptability of 2450 MHz to the uniform heating of smaller products such as plated meals could offer advantages that overcome the cost differential.

A disadvantage of 915 MHz strangely ensues from its major advantage, large module size. Essentially no low-power equipment is available for experimental prototype testing! It is thus very difficult to determine a priori what heating patterns may be at 915 MHz. It is very unwise to do prototype testing at 2450 MHz and assume similar results exist in a production 915-MHz system.

TABLE 10-4   Comparison of Processing Attributes at 915 MHz and 2450 MHz

Attribute	915 MHz	2450 MHz
Generator power	Typical generator size: 30 and 60kW. 10 kW module developed in China.	Module size: 7 kW. 10 kW modules under development.
System cost	Generators: $1.00–$1.50 per watt. Conveyor $50,000 to $150,000 depending on complexity. No sterilization system developed. Does not include installation.	Generators: $2.25 per watt. Conveyor system: $50,000 to $150,000 depending on complexity. Sterilization systems more. Does not include installation.
Maintenance	Line shut down when magnetron fails on medium power systems (50-100 kW). Only 20-min changeover time required.	Several magnetron failures do not impair line operation. Large number of magnetrons may require excessive maintenance.
Heating	Dielectric parameters similar at both frequencies due to compensating contribution of ionic and dipolar effects (Figure 5-4). Penetration depth greater at 915 MHz due to lower frequency. Heating uniformity in general easier to achieve at 915 MHz. Excels for tempering.	Amenable to microwave-field-shaping applicators providing uniform heating of plated meals.

Several years ago a 10-kW, 915-MHz magnetron was introduced by the People's Republic of China. This tube would make an excellent, modest-cost prototype system but has not yet received widespread attention. An excellent entrepreneural situation exists for the establishment of a testing laboratory.

At this point, if cost estimates appear acceptable, a more detailed economic analysis may be undertaken (Jolly 1976), or contracting for further microwave testing should be undertaken. Microwave processing equipment manufacturers are listed in Appendix 4.

CONCLUSIONS

Microwave processing is not the panacea that one might assume from the highly visible success of the microwave oven. There are, however, opportunities for microwave use if attention is paid to their selection. The following list is a guideline for whether or not microwave processing may be a viable choice.

1. Consider bulk drying or dehydration with caution. Most applications are not economically viable.
2. If space is at a premium and increased throughput is required, consider augmenting a present system in either the increasing- or falling-rate region.
3. Most successful product candidates come from those that have high intrinsic value. Economic considerations usually eliminate commodity products from consideration.
4. Applications of microwaves to freeze or vacuum drying so far are rare. Consider opportunities where the product value is high or cannot be produced by any means other than with microwaves.
5. Blanching might have unique applications where products are sensitive to conventional treatment. Increased organoleptic quality might overcome cost considerations.
6. The success of continuous microwave vulcanization of rubber could point to excellent opportunities for continuous processing of extruded food products, particularly those of high value.
7. Probability of high success could come from the combined use of microwaves with other forms of processing. The potential synergistic effects of microwaves combined with steam, forced-air convection, and/or infrared will probably lead the future expansion of microwave processing technology.

References

Copson, D. 1962 and second edition 1975. *Microwave Heating.* Westport, CT: AVI (out of print).

Decareau, R. 1985. *Microwaves in the Food Processing Industry.* New York: Academic Press.

Decareau, R. 1988. Emerging status: shelf-life extension potential of microwave pasteurization. In *Proceedings of the 2nd Annual Conference on Extended Shelf Life Packaging for Foods.* The Packaging Group. East Brunswick, NJ.

Decareau, R. 1990. Microwave uses in food processing. In *Food Technology International Europe 1990.* London: Sterling.

Decareau, R. and Peterson, R. 1986. *Microwave Processing and Engineering.* Chichester, England: Ellis Horwood.

Earle, R. 1983. *Unit Operations in Food Processing.* New York: Pergamon Press.

FCC 1991. Federal Communications Commission. "Industrial, Scientific, and Medical Equipment," Code of Federal Regulations, Title 47, Part 18.

Jolly, J. 1976. Economics and energy utilization aspects of the application of microwaves: A tutorial review. *Journal of Microwave Power* 11(3):233–245,

Metaxas, A. and Meredith, R. 1983. *Industrial Microwave Heating.* London: Peter Peregrinus.

Mudgett, R. 1982. Electrical properties of foods in microwave processing. *Food Technology* **February:**109–115.

Ohlsson, T. 1987. Sterilization of foods by microwaves. Read at International Seminar on New Trends in Aseptic Processing and Packaging of Foodstuffs, Munich. (Available from SIK; Report 1989 nr 564.)

Ohlsson, T. 1989. Dielectric properties and microwave processing. In *Food Properties and Computer-Aided Engineering of Food Processing Systems,* (R. Singh and A. Medina, eds.), pp 73–92. Boston: Kluwer Academic. (Available from SIK; Report 1989 nr 494.)

O'Meara, J. 1973. Why did they fail? (A backward look at microwave applications in the food industry). *Journal of Microwave Power* 8(2):167–172.

Osepchuk, J. M. 1984. A history of microwave heating applications. *IEEE Transactions on Microwave Theory and Techniques* 32(9):1200–1224.

Puschner, H. 1966. *Heating with Microwaves.* New York: Springer-Verlag. (out of print).

SI PLUS 1989; Geocomp Corp. Concord, MA.

Stuchly, M. and Stuchly, S. 1983. Industrial, scientific, medical and domestic applications of microwaves. *IEE Proceedings* 130(A8):467–503 (copious references).

Smith, R. 1984. *Microwave Power in Industry.* Palo Alto, CA: Electric Power Research Institute, Report EM-3645.

Wilson, T.M. 1987. *Radio-Frequency Dielectric Heating in Industry.* Palo Alto, CA: Electric Power Research Institute Report EM-4949.

Zimmerly, J. 1991. Private communication. PSC, Inc. Cleveland, OH.

Glossary

(Commonly used terms not defined in the text)

Agar   A gel made from marine algae and usually used as a bacterial culture medium. With salt added, can be used to simulate the dielectric properties of foods.

Algorithm   A set of rules used to perform calculations or operations. The rule set used by a microprocessor oven controller to perform its controller operation.

Amplitude   The magnitude of a varying quantity such as a voltage or electric field (cf. Figure 2-7).

Amplitude modulation   The variation of the amplitude of an electromagnetic signal corresponding to voice or other information to be transmitted.

Blue Sky   The concept of gathering a group of people together to freely propose ideas for the solution of a problem.

Bumping   A term applied to viscous products that "bump" or "burp" when heated in a microwave oven due to entrapped steam pockets.

Chip   See microprocessor.

Decibel (dB)   A logarithmic representation of a ratio of two fields E_1/E_2, or powers, P_1/P_2, used in electronics because ratios expressed as decibels are added instead of multiplied (one tenth of a bel).

$$dB = 20 \log_{10} \frac{E_1}{E_2}$$

or

$$dB = 10 \log_{10} \frac{P_1}{P_2}$$

Electrode   An electronic conductor, often a metal plate, used to collect or emit electrons. Used in batteries and vacuum tubes.

Emissivity   The effectivity of a body to emit or absorb electromagnetic radiation; commonly used to describe infrared properties. Blackbodies or dark bodies have emissivities near 1 (100% effective), while light or reflective bodies have low emissivities.

Far Field   A propagating electromagnetic wave far from its generator. In this region the electric and magnetic field components of the wave are orthogonal (cf. Figure 1-5).

Ferrite   A magnetic ceramic material containing iron, used for permanent magnets and microwave absorbers.

Frequency Modulation   The variation of the frequency of an electromagnetic signal corresponding to voice or other information to be transmitted.

Heat Capacity   The amount of energy required to raise a unit mass of a material one degree. Units typically are J/kg K, cal/gm°C, and BTU/lb°F.

Histogram   A graphical represention, using vertical bars, representing the number of items within a particular range of properties. For example, a histogram may give the number of microwave ovens sold in a country in a year with power output between 400 and 500 W, 500 and 600 W, etc.

Impedance   The opposition to the flow of current in an alternating current circuit given by the ratio of voltage applied to current flowing. For an electromagnetic wave, the impedance is the ratio of the electric field to the magnetic field in the propagation medium.

ISM   Industrial, scientific, and medical. Electromagnetic spectrum frequency bands allocated for use for these purposes (cf. Table 10-2).

Liquid Crystal   Liquids whose molecules are aligned in a regular pattern having electrically controllable anisotropic properties. Used for digital displays in microwave ovens, watches, etc.

Lossy   Efficient in absorbing electromagnetic or microwave energy and converting it to heat.

Microprocessor (chip)   A small semiconductor circuit containing numerous electronic devices, such as transistors, that perform complex calculations; mainly used in computers. The entire algorithm for a microwave oven controller can be incorporated on a single chip.

Microwavable (often misspelled microwaveable)   Able to be heated, cooked, or prepared in a microwave oven.

Near Field   An electromagnetic wave near its generator. Electric and magnetic fields may have various configurations and are usually not orthogonal (cf. Figure 1–5).

Organoleptic   Pertaining to characteristics as perceived by the sense organs. Used to describe the composite quality of a food's smell, taste, appearance, and mouth feel.

Orthogonal   Perpendicular to or at right angles to each other such as the electric and magnetic fields in a propagating plane wave.

Penetration   A marketing term indicating the number or percentage of a product sold in a given market. This term is often confused with saturation.

Phantom   A material used to mimic another. Especially used as a substitute for food products in microwave testing.

Plasma   A gas whose atoms or molecules have been ionized and is characterized by a colored glow such as seen in a neon bulb. Often occurs under low pressure in microwave freeze-drying experiments.

Proximate Analysis   The approximate composition of a food product consisting of its proportions of water, carbohydrate, protein, fat, and ash.

psia   Pounds per square inch, absolute. A measure of absolute pressure. For example, 14.7 psia is the approximate pressure at sea level on the earth (cf. Table 10–3).

psig   Pounds per square inch, gauge. A measure of pressure referenced to atmospheric pressure. Usually the pressure measured on a pressure gauge. For example, 0 psig is the pressure measured at sea level on the earth (cf. Table 10–3).

Reflection Coefficient   The ratio of the amplitude (electric field) of a reflected wave to its incident value when impinging on a surface. (Often confused with the fractional reflected power, P_r, Equation 5–2).

Retorted   Sterilized or canned by processing in a pressurized retort or vessel in order to raise the temperature to 121°C to kill microorganisms.

rms   Root mean square; the square root of the arithmetic mean of the squares of a set of numbers. When applied to an alternating sine-wave current, the rms value represents the equivalent heating capability compared to a direct current of the same magnitude. The rms value of a sine wave is 0.707 times its peak value.

Saturation   A marketing term representing the total predicted percentage or number of a product that it is possible to sell in a given market at some time in the future. This term is often misused in place of penetration.

Shelf Stable   A food product that has been retorted so that it requires no refrigeration for storage. Product may be in can, pouch, or sealed tray. Recommended shelf life for shelf-stable products is approximately one year.

Sine Wave   A continuous variation of a property, usually an amplitude, from zero, through a maximum, returning to zero, through a maximum in the opposite direction, and finally returning to zero (cf. Figure 1-3).

Sinusoidal   Alternating in the form of a sine wave.

Spores   A microorganism in a dormant state or a one-celled reproductive organ of a fungus. Spores may be activated by appropriate environmental conditions.

Specific Heat   Ratio of heat capacity of a material to that of water. Often used in place of the more correct term heat capacity.

VSWR   Voltage standing-wave ratio. The ratio of the maximum amplitude of an electromagnetic wave to its minimum amplitude. VSWR is commonly used as a measure of the reflected component of a wave. An infinite VSWR means that 100% of the wave has been reflected. A VSWR of 1 indicates that none of the wave has been reflected.

FUNDAMENTAL PROPERTIES OF FREE SPACE

Permittivity

$$\epsilon_0 = 10^{-9}/36\pi \text{ F/m}$$

Permeability

$$\mu_0 = 4\pi \times 10^{-7} \text{ H/m}$$

Velocity of propagation of electromagnetic wave

$$c = 2.998 \times 10^8 \text{ m/s} = \frac{1}{\sqrt{\mu_0 \epsilon_0}}$$

Impedance

$$Z_O = 377 \text{ ohms} = \frac{\mu_0}{\epsilon_0}$$

A number of definitions have been adapted from *The American Heritage Dictionary of the English Language.* ed. W. Morris; Boston, Houghton Mifflin Company, 1981.

Note: For a more comprehensive list of technical terms the following references are suggested:

Microwave Cooking Handbook. Clifton, VA: International Microwave Power Institute, 1987.

IEEE Dictionary of Electrical and Electronic Terms. IEEE Standard 100-1988. ed. C. Booth. New York: IEEE Press, 1988.

Appendix 1

Use of Common Prefixes in Microwave Technology (From Units and Conversion Charts. New York: IEEE Press, 1991)

Prefix	Multiplier	Prefix	Prefix Applied to Hertz	
none	1		Hz	Hertz
kilo	1,000	k	kHz	Kilohertz
mega	1,000,000	M	MHz	Megahertz
giga	1,000,000,000	G	GHz	Gigahertz
tera	10^{12}	T	THz	Terahertz*
peta	10^{15}	P	PHz	Petahertz*
eta	10^{18}	E	EHz	Exahertz*
milli	10^{-3}	m	ms	Millisecond
micro	10^{-6}	μ	μs	Microsecond
nano	10^{-9}	n	ns	Nanosecond
pico	10^{-12}	p	ps	Picosecond
femto	10^{-15}	f	fs	Femtosecond*
atto	10^{-18}	a	as	Attosecond*

*Rarely used.

Appendix 2

Definitions of Dielectric Properties

ϵ_{abs} absolute complex permittivity (often called absolute complex dielectric constant).

$$\epsilon_{abs} = \epsilon'_{abs} - j\epsilon''_{abs}$$

ϵ'_{abs} = absolute permittivity
ϵ''_{abs} = absolute dielectric loss factor

ϵ relative complex permittivity or ratio of absolute permittivity to that of free space, i.e. vacuum (often called relative *complex dielectric constant*).

$$\epsilon = \frac{\epsilon_{abs}}{\epsilon_0} = \frac{\epsilon'_{abs}}{\epsilon_0} - j\frac{\epsilon''_{abs}}{\epsilon_0}$$

where ϵ_0 is the permittivity of free space
$\epsilon_0 = 10^{-9}/36$ F/m, International Scientific (SI) units
$\epsilon_0 = 1$, centimeter, gram, seconds (cgs) units

Thus, $\epsilon = \epsilon' - j\epsilon''$
ϵ' = relative real permittivity (often called relative real *dielectric constant*)
ϵ'' = relative *dielectric loss factor*

NOTES

1. An asterisk is occasionally used to denote complex values, e.g., μ^* or ϵ^*. International consensus, however, does not recommend its use.

    When complex mathematical notation is used, j is defined as $\sqrt{-1}$.

2. The subscript r is occasionally used to denote relative values. Dropping the r provides a simpler notation and is preferred.

3. Terms given in italics are in common usage throughout the microwave cooking and processing industries. These abbreviated forms (italicized) are used throughout the text but should be thought of in terms of their full definitions.

4. The term *permittivity* is preferred in microwave engineering. Dielectric constant is more frequently used in food and material science.

5. Occasionally $\kappa = \kappa' - j\kappa''$ is used for relative permittivity, with ϵ's used for absolute permittivity and permittivity of free space. κ was popularized by the Dielectric Laboratory at MIT in the 1950s. At the time, most literature in the field used the cgs system of units, where the absolute and relative permittivities are equivalent; thus ϵ became used interchangeably with κ. With the adoption of SI units the distinction between the two needed to be clarified, and the use of κ has all but disappeared.

Appendix 3

Dielectric and Thermal Properties of Foods and Materials* (2450 MHz and 20–25°C, except where noted)

*Data extracted from the references of Chapter 5.

Food or Material	Dielectric Constant ϵ'	Loss Factor ϵ''	Penetration Depth d_p (cm)	Density (kg/m³)	Heat Capacity C_p (J/kg K)	Thermal Conductivity k (W/m K)	Thermal Diffusivity (m²/s)	Remarks
Distilled water	77.4	9.2	1.7	1000	4180	0.6	1.4×10^{-7}	24°C
Water + 1% NaCl	77.1	23.6	0.73	1005	4100			24°C
Water + 5% NaCl	67.5	71.1	0.25	1034	3725			24°C
Ice	3.2	0.003	1162	920	2090	2.25	11.7×10^{-7}	0°C
Vegetables								
Potatoes (raw)	62	16.7	0.93	950		0.55		
Peas (cooked)	63.2	15.8	1.0					
Carrots (cooked)	71.5	17.9	0.93					
Vegetable soup	70	17.5	0.94					
Fish, cod (cooked)	46.5	12	1.13	720		0.5		
Fruit (raw)								
Banana	61.8	16.7	0.93	930	3350			
Peach	71.3	12.7	12.7	930	3770			
Meat								
Beef (lean, raw)	50.8	16	0.87	1080	3600	0.5	1.3×10^{-7}	
Beef (cooked)	35.4	11.6	1.0					
Beef (cooked)	32.1	10.6	1.1					60°C
Turkey (cooked)	39.	16.	0.8		3810			

Pork (lean, raw)	53.2	15.7	0.9	1050	3800	0.5	1.3×10^{-7}
Ham	57.4	33.2	0.46		2350		
Ham	85	67	0.3				
Cooking oil	2.5	0.1	23.7	910	2010	0.17	0.9×10^{-7} 60°C
Cooking oil	2.6	0.2	19.5	900	2010		60°C
Butter (salted)	4.4	0.5	8.2		2010		
Butter (unsalted)	3.	0.1	30.5		2010		
Gravy	73.4	26.4	0.64	~1000	3345		
Catsup	54	40	0.36				
Mustard	56	28	0.52				
Bread	4	2	2				
Materials							
Fused quartz	4	0.005	1170	2200	710	1.5	10.2×10^{-7}
Borosilicate glass	4.3	0.02	200	2250	837	1.13	6.0×10^{-7}
Glass ceramic	6	0.03	160				
Soda lime glass	6	0.1	40	2600	670	0.88	5.1×10^{-7}
Teflon®*	2.1	0.0006	4920	2100		0.2	
Thermoset polyester	4	0.02	195				
Nylon	2.4	0.02	150				
Paper	3–4	0.05–0.1	50				
Wood	1.2–5	0.02–0.5	115	500–820	1590–1800	0.08–.36	1.5×10^{-7}

Notes: Where thermal properties are not available, estimates can be made by referring to the techniques described in Chapter 6 and Equations 6-2.
*A registered Trademark of DuPont

Appendix 4

Microwave Resources: Equipment Manufacturers, Laboratories, and Consultants

Facility	Resources Available
A. D. Tech Miles Standish Industrial Park Taunton, MA 02780	Custom and production susceptor film
AGEMA Infrared Systems 550 County Ave. Secaucus, NJ 07094	Infrared camera and imaging systems
Associated Sciences Research Foundation 126 Water Street Marlborough, NH 03455	Technical consulting, equipment design laboratory generators, waveguide components, microwave courses
Atlantic Microwave Route 117 Bolton, MA 01740	Waveguide components
Ball Corporation 3400 Gilchrist Road Mogadore, OH 44260	Production temperature monitoring probes and systems
Berstorff An Der Breiten Wiese 3 W-3000 Hannover 61, Germany	Industrial microwave equipment, pasteurizing systems
CEM 3100 Smith Farm Road Matthews, NC 28106	Microwave chemical systems

Facility	Resources Available
Center for Professional Advancement 46 W. Ferris, Box H East Brunswick, NJ 08816	Microwave educational courses
Cober Electronics 102 Hamilton Avenue Stamford, CT 06902	Industrial microwave equipment, laboratory instrumentation
Decareau, Robert V. Box 241 Amherst, NH 03031	Consulting: food processing, food service, and food product development
Enersyst Industrial 2051 Valley View Ln. Dallas, TX 75234	Microwave design engineering services, microwave-convection systems
Fluke Box 9090 Everett, WA 98206	Voltmeters, multimeters, test equipment
Gamma Consultants 1 Montague Terrace Durham Road Bromley, Kent, UK BR2 0SZ	Microwave consulting, market analysis, oven design
Geocomp 66 Commonwealth Ave. Concord, MA 01742	Scientific units conversion software
Gerling Laboratories 1132 Doker Dr., #1 Modesto, CA 95353	Microwave test and waveguide equipment, consulting
Hewlett Packard 1400 Fountaingrove Pkwy. Santa Rosa, CA 95403	Dielectric measurement equipment, network analyzers, microwave test equipment
HITM 830 Transfer Rd., Suite 35 St. Paul, MN 55114	Marketing and consulting services, custom course preparation, food safety
Holaday Industries 14825 Martin Drive Eden Prairie, MN 55344	Microwave leakage detection equipment
International Microwave Power Institute (IMPI) 13542 Union Village Circle Clifton, VA 22024	Professional microwave organization serving consumer and industrial applications, microwave technology resources

Facility	Resources Available
Ircon 7301 North Caldwell Ave. Niles, IL 60648	Infrared temperature measurement systems
James River 8044 Montgomery Cincinnati, OH 45236	Susceptor boards
Keefer, Richard 630 Bolivar Street Peterborough, Ontario Canada K9J 4S2	Microwave consulting services
Lectronics 1423 Ferry Ave. Camden, NJ 08104	Used microwave and test equipment; obsolete equipment manuals
Lehighton Electronics Box 328 Lehighton, PA 18235	Susceptor resistivity measurement instrumentation
Luxtron 1060 Terra Bella Ave. Mountain View, CA 94043	Microwave-compatible fiber-optic temperature and electric field measurement equipment
Microwave Heating Heron Trading Estate, #2 Sundon Park, Luton, Beds England LU3 3BB	Industrial microwave equipment, sterilization and pasteurizing systems
Metricore 18800 142nd Ave. Mountain View, CA 98072	Microwave compatible fiber optic temperature and pressure measurement equipment
Microdry 7450 Highway 329 Crestwood, KY 40014	Industrial microwave equipment
Microtrans Box 7 S-43801 Landvetter, Sweden	Microwave engineering consulting
Mudgett, Richard Amherst, MA 01003	Microwave modeling consulting
Narda Microwave 435 Moreland Road Hauppauge, NY 11788	Microwave leakage detection equipment

Facility	Resources Available
Oil Center Research Box 51871 Point des Mouton Road Lafayette, LA 70507	Phantom dielectric material
Omac Via Industria 6 Pratissolo di Scandiano, RD Italy 42019	Industrial sterilization and pasteurization equipment
Omega 1 Omega Drive Stamford, CT 06907	Temperature monitoring equipment, data loggers and recorders
PSC 21761 Tungston Road Cleveland, OH 44117	Radio-frequency processing equipment
Richardson Electronics 40W 267 Keslinger Rd. LaFox, IL 60147	Microwave tubes, magnetrons
Rubbright Group Box 11297 St. Paul, MN 55111	Marketing consulting, competitive analysis, product testing and evaluation
Schiffmann, R. F. & Associates 149 West 88th St. New York, NY 10024	Consulting: food products, process and industrial systems R&D, packaging and utensil design, product testing, evaluation recipe design, seminars
Schrade Hochfrequenztechnik Ferdinand Gabriel Weg 12 4770 Soest, Germany	Microwave pasteurization systems
SIK (Swedish Food Institute) Box 5401 S-40229 Göteborg, Sweden	Contract consulting, multiclient studies
Steyskal, H. & Associates 585 Harrington Ave. Concord, MA 01742	Electromagnetic theory consulting services
Surhart Corporation 177 Frank Street Ottawa, Canada K2P 0X4	Dielectric measurement software, consulting

Facility	Resources Available
Raytheon Corporation Foundry Ave. Waltham, MA 02154	Industrial microwave equipment, microwave tempering systems
Technipower 14 Commerce Dr. Danbury, CT 06810	Line voltage controllers Power monitors
Wiltron 490 Jarvis Drive Morgan Hill, CA 95037	Microwave measurement instrumentation, network analyzers

The International Microwave Power Institute (IMPI) serves as a data bank for all types of microwave-related resources. Also refer to IMPI *Microwave Reference Guide.*

For a more complete listing of worldwide suppliers of microwave processing equipment see *Microwaves and Food Newsletter,* January 1992, published by Food and Nutrition Press, Trumbull, CT.

Appendix 5

Power Measurement Test Procedures

IMPI 2-Liter Test

(Adapted from Buffler, C. 1991. A guideline for power output measurement of consumer microwave ovens. *Microwave World* **10**(5):15.)

Operate the oven at its rated line voltage with oven set on high with a load of 2000 ± 5 g placed in two 1-L beakers such as Pyrex 1000 or Kimax 1400. The beakers should initially be at room ambient temperature. Initial water temperature should be 20°C ± 2°C (64.5–71.5°F), measured after water is placed in beakers and before placing in the microwave oven. The beakers are placed in the center of the oven, side by side in the width dimension of the cavity, and touching each other. The oven is turned on for 2 min and 2 s. The beakers are removed from the oven, and the final temperatures are measured and recorded.

The power is calculated from the following formula:

$$P \text{ (W)} = 70 \times \frac{\Delta T_1 \, (°C) + \Delta T_2 \, (°C)}{2}$$

where ΔT_1 and ΔT_2 are the temperature rises of the water in the two beakers, calculated by subtracting the initial water temperature from the final temperature.

The power measurement should be run three times, with the oven power the average of the three readings. If any individual measurement is more than 5% from the average, the complete test should be repeated.

Note: The oven should be prewarmed by heating 2L of water for 5 minutes, then wiping the shelf with a cold wet rag (c.f. Chapter 8).

Note: The water in each vessel should be well stirred before measuring the starting and final temperatures. A small object, such as a plastic spoon or the handle of a wooden spoon, should be used. Do not use metal! The temperature should be measured with a thermometer or thermocouple with 0.1°C resolution. Such instruments may be obtained from chemical supply houses.

IEC Test

(Adapted from *Methods for Measuring the Performance of Microwave Ovens for Household and Similar Purposes,* 2nd ed. CEI IEC 705. (Geneva, Switzerland: Bureau Central de la Commission Electrotechnique Internationale.)

Operate the oven at its rated line voltage with oven set on high with a load of 1000 ± 5 g placed in a 190-mm-diameter, 1-L vessel such as a Pyrex 190 × 100 #3140 crystallizing dish. The vessel should initially be at room ambient temperature. Initial water temperature should be 10°C ± 2°C (46.4–53.6°F), measured immediately before the water is placed in the vessel and before placing in the microwave oven. The vessel is then placed in the center of the oven.

By trial and error, the time *t,* in seconds, required to raise the temperature of the water 10°C ± 2°C is determined. (Repeated trials may be required. Approximately 1 min may be required for a 700-W oven.) The water should be well stirred before measuring the starting and final temperature. A small object, such as a plastic spoon or the handle of a wooden spoon, should be used. Do not use metal! The temperature should be measured with a thermometer or thermocouple with 0.1°C resolution. Such instruments may be obtained from chemical supply houses.

The power is calculated from the formula

$$P \text{ (W)} = \frac{4187 \times \Delta T \text{ (°C)}}{t \text{ (s)}}$$

where ΔT is the temperature rise of the water in the vessel, calculated by subtracting the initial water temperature from the final temperature.

The power measurement should be run three times, with the oven power the average of the three readings. If any individual measurement is more than 5% from the average, the complete test should be repeated.

Consumer One-Cup Test

(Adapted from Buffler, C. 1991. Standards committee report. *Microwave World.* **12**(4):10.)

This practical test is designed to use a one-cup (U.S.), 237-milliliter measure.

From a container of half ice and half water, measure exactly one cup of water (no ice) into a glass measuring cup. Place in center of microwave oven shelf. Heat on high for 5 min or until water begins to boil. If water begins to boil in less than $3\frac{1}{2}$ min, consider your oven high power; if longer, the oven is low power.

Note: For a full explanation of the details of each test, refer to the cited references. For derivation of the power formulas, refer to Chapter 4.

Appendix 6

Common Waveguide Characteristics

Waveguide Designation WR-	Frequency Range (GHz)	Cutoff Frequency (GHz)	Guide Wavelength* in(cm)	Inside Dimensions in(cm)	Wall Thickness in(cm)	Attenuation (approximate; db/100 ft)
975	0.75–1.12	0.605	17.20 (43.69)	9.75 × 4.875 (24.77 × 12.38)	0.125 (0.318)	0.1
430	1.7–2.6	1.372	5.82 (14.78)	4.30 × 2.15 (10.92 × 5.46)	0.080 (0.203)	0.3
340	2.2–3.3	1.736	6.83 (17.35)	3.40 × 1.70 (8.64 × 4.32)	0.080 (0.203)	0.8
284	2.6–3.95	2.078	9.09 (23.09)	2.84 × 1.84 (7.21 × 4.67)	0.080 (0.203)	1.0

Notes:

1. The velocity of light used for the calculations was the more exact value of 2.998×10^8 m/s.
2. Waveguide is commercially available in aluminum or copper alloy (brass). Aluminum waveguide has slightly less attenuation than brass.
3. Attenuation of waveguide increases rapidly toward infinity between lower end of the designated frequency range and the cutoff frequency.
4. A second, higher-order, mode is able to propagate above the upper end of the designated frequency range.
* Guide wavelength is designated for 0.915 GHz in WR-975 waveguide and 2.45 GHz for WR-430, 340, and 284 waveguide.

Index